NEW ENGLAND INSTITUTE
OF TECHNOLOGY
LEARNING RESOURCES CENTER

Commercial Cool Storage
Design Guide

COMMERCIAL COOL STORAGE DESIGN GUIDE

Electric Power Research Institute
EM-3981
Research Project 2036-3

● **HEMISPHERE PUBLISHING CORPORATION**
A Subsidiary of Harper & Row, Publishers, Inc.
Washington   New York   London

Distribution Outside North America
SPRINGER-VERLAG
Berlin   Heidelberg   New York   London   Paris   Tokyo

15223492

10-89

Commercial Cool Storage Design Guide

Library of Congress Cataloging-in-Publication Data

Main entry under title:

Commercial cool storage design guide.

   "EM-3981."
   "Research project 2036-3."
   Bibliography: p.
   Includes index.
   1. Cold storage — Energy conservation.   2. Electric power consumption.   I. Electric Power Research Institute.
TP372.2.C65 1987   664'.02852   87-136
ISBN 0-89116-687-4   Hemisphere Publishing Corporation

DISTRIBUTION OUTSIDE NORTH AMERICA:
ISBN: 3-540-17715-9   Springer-Verlag Berlin

CONTENTS

PREFACE

Cool storage — a technique for shifting all or part of the air conditioning requirements from peak to off-peak hours — offers the potential to reduce peak electricity demands and generate significant savings in electric bills. Several hundred cool storage installations already exist in the United States and Canada. The technology is particularly attractive to utilities in summer-peaking areas, where it can help control demand growth and improve the use of available generating capacity. A number of these utilities now offer special rates, incentives, and other programs designed to promote cool storage in their service territories.

One obstacle to greater acceptance has been the lack of information about system alternatives and design procedures. It is the purpose of this handbook to provide comprehensive guidance for designing ice and chilled-water storage systems for commercial buildings.

The guide contains or references state-of-the-art information necessary to evaluate the cost-effectiveness of cool storage options; select, configure, and screen system alternatives; and carry out heating, ventilating, and air conditioning (HVAC) system design incorporating cool storage.

Both chilled-water and ice storage systems are described, including techniques and design data for feasibility analysis, cost-effective system sizing and operation, design of storage tanks, selection of refrigeration components, design of water and air distribution systems, and equipment maintenance. Deviations from conventional HVAC design and common design errors are also discussed. Two case studies illustrate successful cool storage applications, and a final section lists information sources, such as manufacturers, trade organizations, and R & D institutions.

ACKNOWLEDGMENTS

We gratefully acknowledge the sponsorship of GPU Service Corporation and Electric Power Research Institute, and the direction provided by George Reeves, Glenn Steiger and Jack Gruitt. We also wish to extend our appreciation to Veronika Rabl for her suggestions and insights.

In addition, we wish to acknowledge the contributions of RCF, Inc. and the Refrigeration Engineering Advisory Council (REAC), William V. Richards in particular. Special thanks also are due to the many organizations and individuals who shared their information and experience. These include the Southern California Edison Company, Ayers Associates, San Diego Gas & Electric, Robert T. Tamblyn, H. C. Yu, and numerous cool storage equipment manufacturers and their representatives.

Chapter 1

INTRODUCTION

Cool storage is an exciting "new" technology first applied more than a half-century ago. Until recent years, however, widespread interest in the concept waned because of the increased efficiency and reliability of conventional cooling systems, and the decreasing cost of electricity. Today's energy situation is such that cool storage has re-emerged as the most advanced, cost-effective space cooling technology of all, particularly due to utility-established customer incentives. These incentives have been created to help utilities operate in a more cost-effective manner.

Most electric utilities experience their peak demand in summer. Meeting this peak demand frequently forces them to rely on less-efficient "peaking plants," and/or to purchase electricity from other grid-connected utilities. In addition, the need to meet peak demand requirements creates more pressure for constructing new generation facilities whose construction and financing costs are far higher than those which were incurred for existing plants, thus creating pressures for higher customer rates.

Virtually all utilities' commercial and industrial rate schedules include a demand charge. Most utilities charge more for demand during the cooling season, and many also employ a "ratchet clause." This requires an owner to pay a demand charge which is at least equal to a percentage of the highest demand recorded for the year, even though actual demand for the billing period involved may have been far less. In addition, some utilities have started to rely on "time-of-use" demand (and energy) rates which impose higher charges for demand recorded during daily periods of utility peak demand, and more utilities are expected to pursue this strategy.

These and other utility policies encourage owners to reduce their buildings' use of electricity during utilities' peak demand periods. In most cases, periods of utility peak demand and buildings' peak demand coincide. In summer, peak demand commonly is caused by space cooling equipment operating to meet cooling requirements.

When cool storage is applied, the time of maximum cooling energy use typically is shifted from periods when the building is already consuming significant amounts of energy to meet occupied-period needs, to off-peak periods when energy use is low. In a typical system, this is accomplished by chilling the storage medium during periods of off-peak building demand. Then, when space cooling is needed during the building's peak periods, the only energy needed to provide it is that associated with fans and pumps.

Although cool storage may not necessarily reduce a building's energy consumption, it will significantly reduce its demand, and thereby create substantial operational cost savings. The value of these savings is enhanced when summer-winter demand rates are used, and particularly when ratchet clauses and/or time-of-use rates are in effect.

Although the demand savings alone often are significant enough to justify reliance on cool storage, the benefits of cool storage go far beyond demand savings alone.

- Operational Flexibility
  Most electric utilities have already implemented or soon will implement demand-side management, an approach which comprises numerous strategies designed to influence utility load shapes in ways which will benefit both the utility and its customers. However, today's load-shaping needs may be far different from those which emerge in five or ten years, creating the possibility of new definitions of utility peak and off-peak periods, new rate schedules, etc. Cool storage provides optimum flexibility in this regard, in that it permits the owners to at all times consume maximum space cooling energy during periods of minimum utility demand.

- More Efficient Equipment Operation
  Conventional systems operate at part-load conditions most of the time. By contrast, cool storage systems run at full-load conditions most of the time, significantly increasing equipment operating efficiency. In addition, because the cool storage system typically operates at night, when outdoor air temperatures are cooler, condenser heat rejection is improved.

- Lower Capital Costs
  Cool storage permits use of a smaller, less costly refrigeration compressor, whose reduced electrical load can result in a lower-cost power distribution system. In addition, when ice is used as the storage medium, lower supply water temperature and a large temperature differential can be employed. This provides an opportunity for lower temperature air distribution with lower CFM, resulting in use of smaller ducts and fan motors.

- Continued Cooling During Power Outages
  In most buildings, conventional cooling is considered a nonessential load and thus is not connected to standby power generation equipment. The relatively small on-peak load imposed by

cool storage can permit such connection, and thus assurance of comfort during a power outage in summer.

- Cost-Saving Utility Incentives
  Some utilities offer special incentives for cool storage, such as cost sharing of feasibility analysis and/or the differential cost (if any) of mechanical equipment. Local utilities should be queried in this regard.

- Better Fire Protection and Lower Insurance Rates
  A stored chilled water system can serve as a supplemental source of water for fire-fighting purposes, enhancing building and occupant safety. This may also result in fire insurance premium reductions

- Increased Building Value and Marketability
  The value and marketability of a building are significantly influenced by the marketability of its HVAC system and the likely future utility costs a new owner would have to pay. In that demand charges will almost assuredly increase in the future, and because time-of-use rates will become more prevalent, the savings created by cool storage annually will continue to increase. As such, cool storage can increase the value and saleability of a building.

- Application Flexibility
  Cool storage systems can be used in new buildings as well as existing buildings, instead of or as a supplement to a conventional cooling system. In some cases, where particularly high demand rates are in effect, and/or where a relatively high cooling load exists for short periods only, it may be cost-effective to replace or at least supplement an existing conventional system with a cool storage system. Reliance on modular cool storage systems permits ease of expansion.

Despite the many benefits of cool storage, some owners may be reluctant to rely on what seems to be a new, relatively untried technology. In this regard, note that one of the first reported installations of cool storage occurred in 1925, when O.L. Moody installed an oversized brine tank in the ice plant of the Baker Hotel (Dallas, TX) (1) to create surplus refrigeration to cool the hotel's coffee shop and cafeteria. Between 1925 and 1936, cool storage systems were installed in six neighborhood theaters and a large retail store in Kansas City, MO (1); in a 50,000-square-foot Kansas City, KS, office building, and in three Houston, motion picture theaters. In 1937, a particularly large cool storage system was installed in a major Dallas, TX (1), hospital, and for many years -- up through the present -- cool storage has been the cooling system of choice for many dairies throughout the United States. Their need for reliable, cost-effective cooling underscores even further the acceptance and "time-tested" aspects of a "new" space cooling method, now applied in the U.S. for more than 60 years.

The variety of cool storage equipment already available makes it feasible to apply the technology to virtually any type of building which ordinarily is billed for demand. Furthermore, the circumstances of today -- and those projected for the future -- make it wise to at least consider the feasibility of using that technology. Its application can create significant initial and life-cycle benefits for the owner and those who use the building, as well as the utility and those others it serves. Most utilities and cool storage equipment manufacturers will be pleased to provide assistance.

PURPOSE

The purpose of this guide is to provide information and techniques for evaluating and designing ice and chilled water cool storage systems, used for space and/or process conditioning applications in buildings. This guide covers various types of cool storage systems that are coming into common use, including full storage which allows cooling equipment to be turned off during certain hours of the day, and partial storage in which cooling is provided by a combination of storage and cooling equipment. It also includes a discussion of the benefits and drawbacks of each approach given specifics of the applications involved; provides guidance relative to cool storage system design procedures; indicates methods available for comprehensive economic analysis, depending on the nature of the utility rate schedules imposed and systems selected, and, finally, presents two case histories of successful cool storage applications.

SCOPE

This guide includes information on ice and chilled water cool storage systems which can be applied to both new and existing commercial, institutional and industrial buildings.

Many types of cool storage systems use storage media other than ice or water, and a variety of experimental technologies and techniques are under development. Nonetheless, this guide is limited to those types of cool storage systems which are operating satisfactorily in a number of buildings.

Mechanical engineers experienced in the design of buildings' heating, ventilating and air conditioning (HVAC) systems comprise the intended audience for this guide.

Chapter 2

COOL STORAGE FUNDAMENTALS

Most buildings are equipped with space cooling systems. Almost all of these systems are electric, and their influence on a building's electrical demand can be costly. Owners of the buildings involved are seeking methods to reduce these expenses without compromising comfort. Cool storage technology is being applied continually more to do just that, helping to improve utilities' load shapes at the same time.

Cool storage systems employ "off-the-shelf" equipment to extract heat from a storage medium during those daily periods when it is most cost-effective to do so (typically during nighttime hours). The cool storage medium is then used to absorb space heat typically during normal occupied daytime hours, using only pump and fan energy in the process. As a result, the building's peak demand is less than it otherwise would be.

Cool storage systems are categorized based on the storage medium used. Ice and water are most commonly employed. Other media -- such as "phase change" materials -- are being developed and tested for suitability.

Cool storage systems also are categorized based on their mode of operation, which determines the size of storage needed. The two basic categorizations commonly are employed together. Accordingly, most cool storage systems are either ice- or water-based full storage or ice- or water-based partial storage. The following discussion first addresses differences between media, and then differences between modes of operation. Before presenting that discussion, it is important to note proper application of the term "peak period."

In most instances, a building's peak demand period occurs during a summer afternoon. By using cool storage, the demand created for cooling is shifted to an off-peak period. However, some utilities employ a time-of-use rate structure which establishes variable pricing for demand, depending on the time of day during which it is created. In these instances, the period of most costly demand may -- or may not -- coincide with that period during which the building records its maximum demand.   Regardless of rate structure, however, "peak period" is that period during which creating electrical demand imposes the most cost. This period almost always coincides with that period during which the utility also experiences maximum demand.

STORAGE MEDIA

The growing popularity of cool storage systems has encouraged a number of manufacturers to investigate development of new types of systems and storage media. The storage medium of one of the newest systems consists of plastic "bricks" filled with a eutectic salt. The bricks are frozen by chilled water from a conventional chiller. Despite the promise of this and other new or emerging cool storage systems, those which use either ice or water as the storage medium remain most popular. Both of these are discussed below, and Table 2-1 comprises a comparison of their most important characteristics. (An in-depth discussion of their comparative advantages and disadvantages is presented in Chapter 3.)

Table 2-1

Chilled Water vs. Ice Storage

ITEMS	CHILLED	ICE
SPACE REQUIRED	LARGE	1/5 - 1/8 OF CHW
POWER REQUIRED	0.70 TO 0.83 KW/TON	0.50 TO 1.3 KW/TON
CHILLED WATER PUMPING POWER	HIGH	LOW
COMPRESSOR SIZES AVAILABLE	VARIOUS SIZES	LIMITED SIZES
MAINTENANCE COST	HIGH	MEDIUM
EXPERIENCE DESIGNER AND OPERATOR	EASILY AVAILABLE	LIMITED

Ice Storage Systems

Ice storage systems are categorized as either static or dynamic. Generally speaking, static systems are smaller, simpler, more efficient, and less costly than dynamic systems. As a consequence, static systems are the more popular of the two types.

Static Systems. The most commonly employed static system is the ice builder, also known as a direct expansion or DX system. Figure 2-1 illustrates a typical ice-

builder system (2). As can be seen, its basic components are similar to those of a
mechanical refrigeration system: compressor, condenser, expansion valve, and a
combination evaporator/thermal storage unit. The evaporator/thermal storage unit
consists of a multiple-tube serpentine coil submerged in a tank of water, and a
water agitation device used to provide uniform ice build-up and melt-down. The
tank is fully insulated and is covered to minimize infiltration of foreign matter.

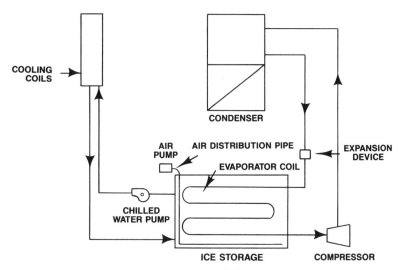

Figure 2-1.   Basic static ice storage system

There are several commercially available ice-builder systems, and all function in
basically the same manner. When fully charged, about half the tank's space is
taken by chilled water, and the other half is taken by ice and evaporator coils.
Supply chilled water is removed from the top of the tank. The return chilled water
is introduced at the bottom of the tank where it contacts the ice built up on the
evaporating coils. The thickness of the ice is controlled to no more than 2" or 3"
to prevent "bridging" and flow blockage. It grows or diminishes depending upon
the relative magnitude of the compressor capacity and the building cooling load.
Uniform ice buildup (and melting) are encouraged either by agitation, e.g., air
bubbles, or by flow path control using baffles.

Static ice-builder systems are available in sizes ranging from 48 to 1,200 ton
hours, and all packaged units can be connected to an existing or new building's
chilled water system. Some are available with a heating circuit which gives them
heat storage capability as well.

Some static systems make ice remote from the evaporator. For example, in the one depicted in Figure 2-2 (3), water is frozen solid around a mat of closely spaced tubes that are rolled up to a vertical position, acting as a heat exchanger. The fluid -- in this case a water/glycol solution -- is circulated through the tubes, entering at 25F and coming out at 32F. The water/glycol solution is pumped from the evaporator to the tanks during the charging cycle, removing heat from the water to cause ice to form. (In this case, the evaporator is one of three major components of a packaged chiller unit. The other two are the compressor and condenser.) The solution then returns to the evaporator through the automatic diverting valve. During the discharge cycle, the ice chills the water/glycol solution which then is pumped through the diverting valve to the duct coil, to cool building supply air. In comparison to an ice-builder, this system requires less storage volume, because it has a higher ice-to-water ratio.

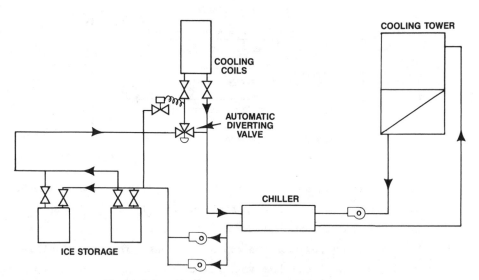

Figure 2-2.   Modular ice storage system using brine

Dynamic Systems. Dynamic systems make ice in plate, chunk, crushed or slush form, and then deliver it for storage in large bins or tanks. An ice slurry system which uses ice in the form of slush is shown in Figure 2-3 (4). When the solution used is cooled to below its freezing temperature, small ice crystals form, creating an ice slurry. The slurry is pumped through the ice generator tube to an insulated storage tank.

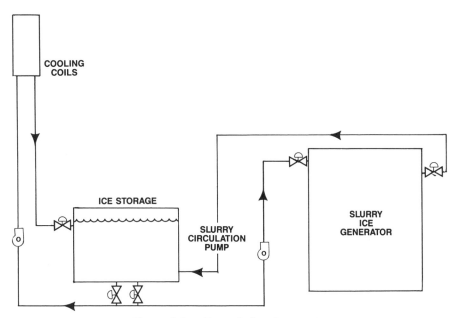

Figure 2-3.   Dynamic ice slurry system

Figure 2-4 illustrates another type of dynamic system, in which water from a storage bin is sprayed over "sprayed freezing coils" (5). Because these coils are external to the storage bin, more bin volume can be filled with ice. This permits use of a smaller freezing surface, but the ice obtained tends to have only half the density of solid ice. The system's freeze cycle is followed by a harvest cycle during which condenser heat is used to free ice from the coils.

Chilled Water Storage Systems

Chilled water storage systems use the heat capacity of water (in its liquid form). Typically they employ a conventional chilled water cooling system augmented by a chilled water storage tank, as shown in Figure 2-5 (3). The factory-assembled refrigeration system, or "chiller", consists of a compressor, condenser and an evaporator. When a water-cooled condenser is employed, a condenser water pumping system and cooling tower (for rejection of condenser heat to the atmosphere) also are needed.

During off-peak periods, the chiller cools water which then is pumped to the water storage tank. When chilled water is needed, it is pumped out of the storage tank to cooling coils, and then is returned to the tank.

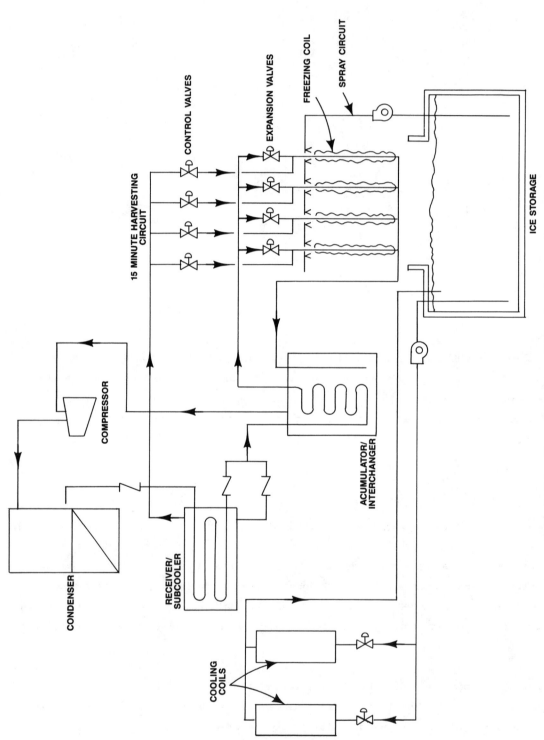

Figure 2-4. Sprayed freezing coil ice-maker storage unit

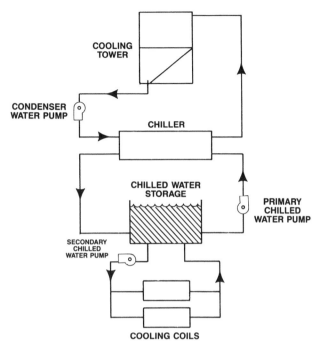

Figure 2-5.   Chilled water storage system

For equivalent cooling capacity, chilled water storage tanks must be far larger than ice storage tanks. Separating chilled water from warmer return water in these tanks is an issue of essential concern, discussed fully in Section 5.

In many applications the storage tanks used for chilled water storage also are employed for storage of hot water produced by using energy during off-peak periods.

OPERATING MODE

The cool storage operating mode desired determines the size of storage capacity required or, conversely, the amount of space available for storage influences the operating mode selected. In either case, a number of variables must be considered, and these are discussed comprehensively in Section 3.

Full Storage

A full storage system provides enough storage to meet a building's full on-peak cooling requirements. The building load profile in Figure 2-6 illustrates how the cooling load increases demand when a conventional cooling system is employed. Figure 2-7 shows how full storage displaces the cooling demand to times when other

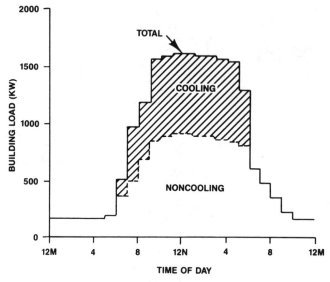

Figure 2-6.   Hourly load profile on the design day

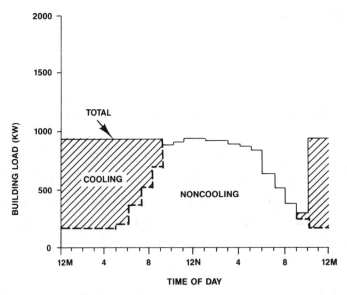

Figure 2-7.   Design day load profile under the full storage operating mode

electrical loads (noncooling loads, e.g., lighting and motors) are negligible and during "shoulder hours," when the other loads start to increase prior to normal building occupancy (6). As a result, all storage cooling occurs during off-peak periods, thus effecting demand cost savings.

Full storage is best suited for applications where on-peak cooling cycle duration is relatively short compared to the time required for ice or chilled water production. In situations in which the building's cooling load is sufficiently large and its daily occupancy period not too long, there will be adequate "room" in the off-peak valley to accommodate the cooling load without causing a nighttime building peak. In situations where there is insufficient room in the off-peak valley, one of the several partial storage operating modes discussed below can be employed.

Full storage systems do not achieve significant reductions in compressor capacity compared to conventional cooling systems. Also, because full on-peak cooling requirements are stored, the storage system must be relatively large.

Partial Storage

A partial storage system runs many more hours than a full storage system so less demand reduction is obtained. However, it is initially less expensive than full storage because less storage capacity is required, and because smaller capacity refrigeration equipment is used.

Partial storage systems can be categorized as either load-levelling or demand-limited. Both types can be configured in several ways. Figure 2-8 illustrates their mode of operation compared to that of conventional cooling systems and full storage, as well as comparative refrigerant plant and storage requirements.

Load-Levelling. When load-levelling cool storage mode is used, capacities of the storage system and refrigeration equipment are selected so the design-day cooling load can be met by continuous operation of the refrigeration equipment. This strategy minimizes compressor capacity requirements and significantly reduces space cooling's contribution to the building's peak demand. As illustrated in Figure 2-9, the overall effect of this operating mode is to level the cooling component of the building's load (6). During peak hours, part of the cooling load is met directly by the compressor and part by storage. The storage required for the partial mode of operation must be adequate to supply all the building cooling load not met directly by the refrigeration equipment. In the situation illustrated in Figure 2-9, about 60% of the building's peak-hour cooling load would be supplied

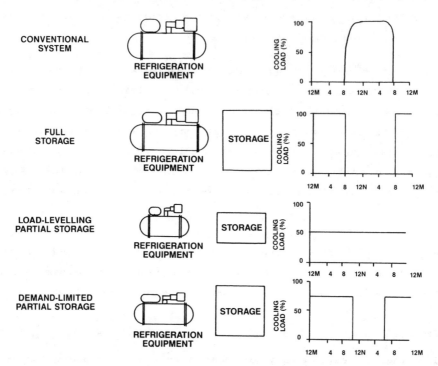

Figure 2-8.   Comparison of cool storage systems

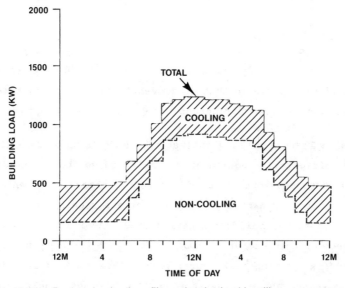

Figure 2-9   Design day load profile under the load-levelling storage operating mode

from storage (<u>6</u>). The fraction met by the compressor increases on either side of the peak until, during "shoulder" hours, compressor output exceeds the direct cooling load and part of the compressor output goes into storage. During off-peak hours, the refrigeration equipment is devoted entirely to cooling the storage medium.

<u>Demand-Limited</u>. Demand-limited systems employ sophisticated techniques to control refrigeration equipment, preventing its operation during peak periods. At all other times, however, the compressor is operated both to meet direct cooling load and to cool the storage medium.

Figure 2-10 illustrates a demand-limited system's effect on a building's load profile (<u>6</u>). As can be seen, the building's peak demand has been reduced to the noncooling peak demand, occurring between 11:00 a.m. and 1:00 p.m. Instead of displacing the cooling load entirely to the off-peak period, as full storage does, a considerable amount of the cooling load is "fit" into the "shoulder" periods, when the noncooling load is "ramping" up or down. The compressor is run during shoulder periods at a level adequate to meet part or all of the direct cooling load, while also cooling the storage medium to that extent which does not cause the total building load to exceed that reached by the noncooling load during peak hours. System operation requires monitoring of the noncooling loads and, for morning "shoulder periods", <u>forecasting</u> the day's peak noncooling demand. In practice, full utilization of the shoulder periods, as illustrated in Figure 2-10, is not likely to be possible.

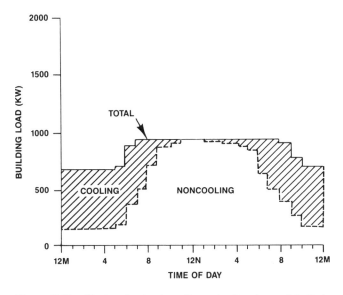

Figure 2-10. Design day load profile under the demand-limited partial storage operating mode

A demand-limited system is most suited for applications where the building's noncooling load is sufficiently large and the daily occupancy period not too long, to provide adequate "room" in the off-peak valley to accommodate the cooling load without causing a nighttime building peak.

Compared to a load-levelling system, a demand-limited system does not achieve as much compressor capacity reduction and requires a larger-capacity storage system. Compressor and storage capacity requirements are less than those of a full storage system, however.

Configurations. Three basic configurations are used with both load-levelling and demand-limited storage systems:

- Compressor-aided,

- Refrigerant coil, and

- Parallel evaporator.

Each of the three is discussed below assuming use of ice as the storage medium.

A compressor-aided configuration is illustrated in Figure 2-11 (2), employing an ice-builder storage system. During the cooling period, the ice charge inside the ice builder acts as an evaporator coil surface, cooling warm return water. Operation of the refrigeration system during this period slows melt-down and thereby provides more cooling capacity than would be obtained in conventional melt-down. This concept eliminates the need for a separate chilled water evaporator or duplicate refrigerant feed equipment, reducing first costs. However, it tends to increase operating costs because the refrigeration system must operate at ice builder evaporator temperatures during melt-down.

An ice storage/refrigerant coil arrangement is shown in Figure 2-12 (2). Ice building occurs during off-peak hours. During on-peak hours (only when a load-levelling system is used), all or part of the refrigeration system capacity circulates refrigerant directly to cooling coils, to meet that portion of the building load not met by melting ice. This arrangement has the lowest initial cost. It does not need an additional chilled water evaporator, and a smaller compressor can be used because refrigerant coils can be operated at a relatively high evaporator temperature (approximately 45F). However, duplicate liquid refrigerant distribution equipment is required for operation at the two different evaporator temperatures, thus increasing the complexity and cost of controls.

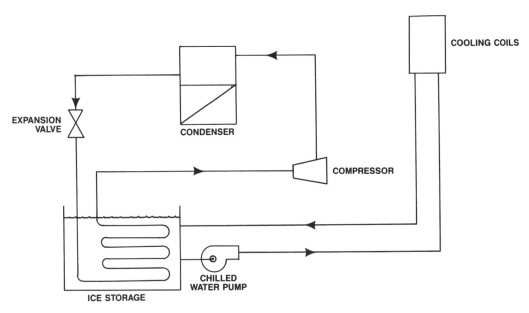

Figure 2-11. Compressor-aided storage system

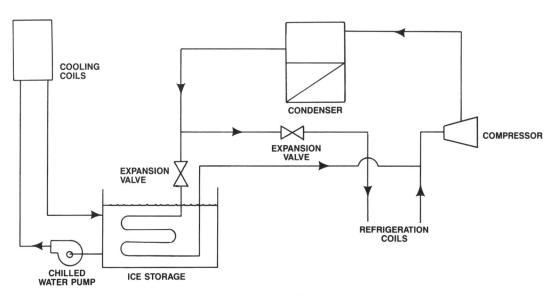

Figure 2-12. Ice storage/refrigerant coil arrangement

The ice storage parallel evaporator arrangement shown in Figure 2-13 uses the same compressor and condenser as the refrigerant coil configuration, but also requires a chilled water evaporator, to supplement ice storage during the cooling period (2). This arrangement has the advantage of using chilled water in all cooling coils, simplifying design. However, the chilled water evaporator required adds substantial initial cost.

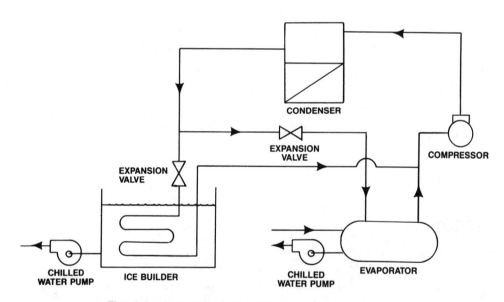

Figure 2-13. Ice Storage/Parallel Evaporator Arrangement

Chapter 3

FEASIBILITY ANALYSIS

A feasibility study generally is performed in order to determine which of several major options is best. Factors considered typically include type of project involved, owner guidelines and job constraints, the equipment's ability to meet estimated performance requirements, economics, other potential benefits and problem areas, ease of maintenance and operation, size and weight, and building codes, among others.

Of these, economics is the most important, particularly because most other factors can be equated in dollar terms. For example, a higher maintenance requirement results in more labor, materials, and spare parts expenses. Likewise, larger size may mean less net rentable or usable area, and more weight may require additional structural bracing.

This section identifies and discusses the ten basic steps (Figure 3-1) associated with performance of a cool storage system feasibility analysis. Although steps are shown in sequential order, some can be performed simultaneously with others.

EVALUATE THRESHOLD CONSIDERATIONS

Certain basic factors will determine the overall feasibility of cool storage, and whether or not it will be worthwhile to perform a feasibility analysis. These factors include utility rate schedules, space availability, and maximum daily integrated load profile.

To be cost-effective, a cool storage system must be able to significantly reduce the cost of demand. For significant savings to result, the cost of demand itself must be substantial. Generally speaking, savings will be particularly sizable when the utility demand charges include a ratchet clause, and/or when time-of-day (TOD) rates are in effect. If demand rates are relatively low and major increases are not forecasted; if there are no ratchet clauses or TOD rates in effect, and none are planned, chances are that cool storage will not be cost-justifiable, making further analysis nonproductive.

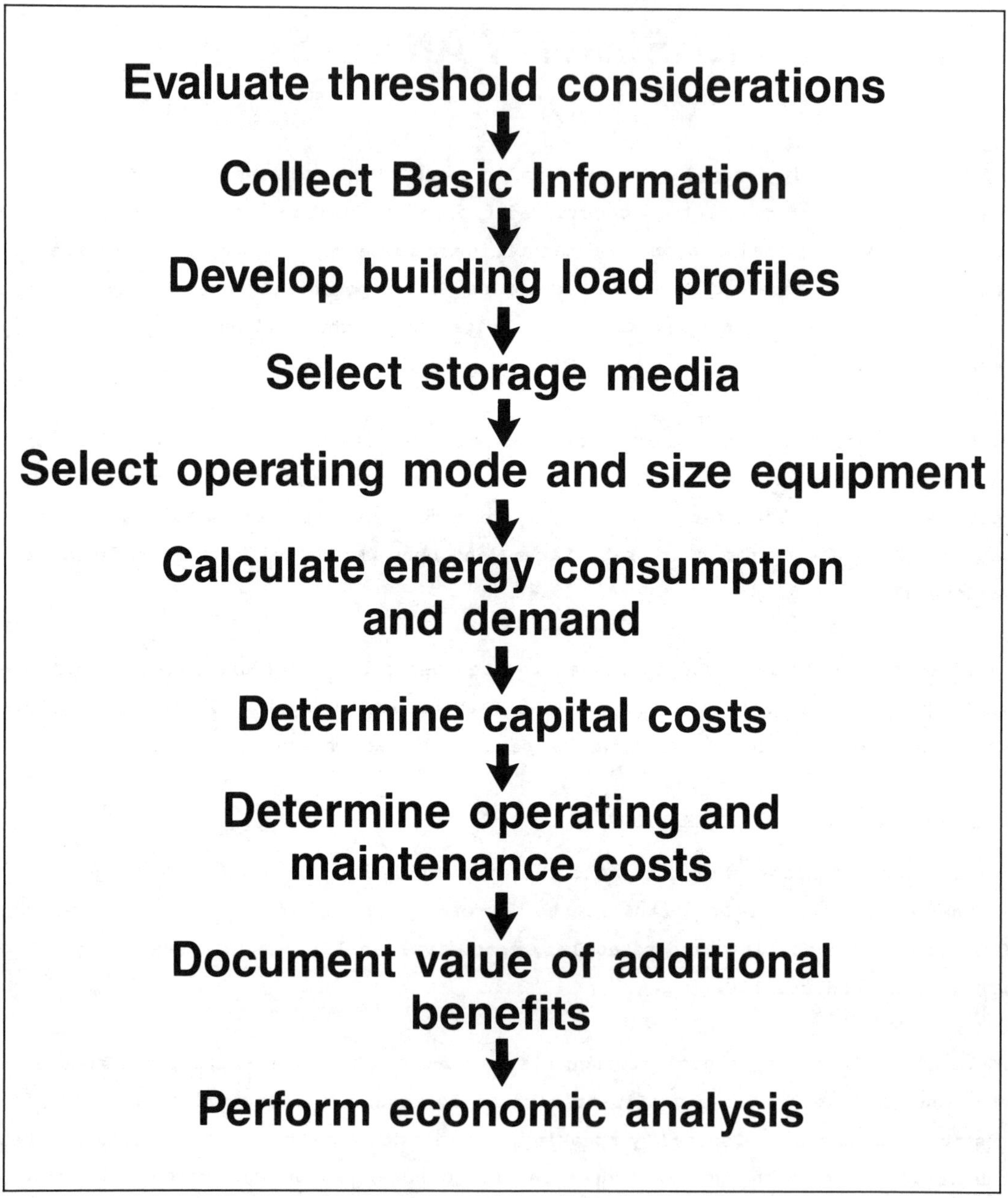

Evaluate threshold considerations
↓
Collect Basic Information
↓
Develop building load profiles
↓
Select storage media
↓
Select operating mode and size equipment
↓
**Calculate energy consumption
and demand**
↓
Determine capital costs
↓
**Determine operating and
maintenance costs**
↓
**Document value of additional
benefits**
↓
Perform economic analysis

Figure 3-1. Steps required for a feasibility study

Space availability for storage tanks is another basic concern, something affected by site conditions, building architecture, and refrigeration plant location.   If storage space cannot be made available, cool storage cannot be applied.

Cool storage is most cost-effective when a high, narrow cooling load profile occurs during the utility's "on-peak" period.   Since storage can reduce the demand peak by moving electrical energy use to off-peak periods, significant utility cost reduction can be achieved.   Types of buildings whose typical load profiles indicate low, medium and high potential for cool storage systems are indicated in Figure 3-2. Figure 3-3 comprises load profiles for office buildings (typically high-potential candidates for cool storage) in four utility service areas (7).

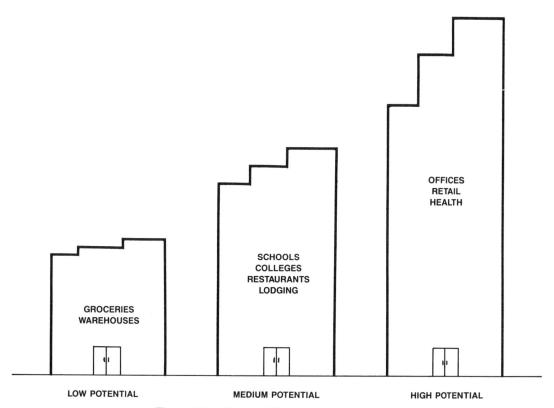

Figure 3-2.   Potential for cool storage systems

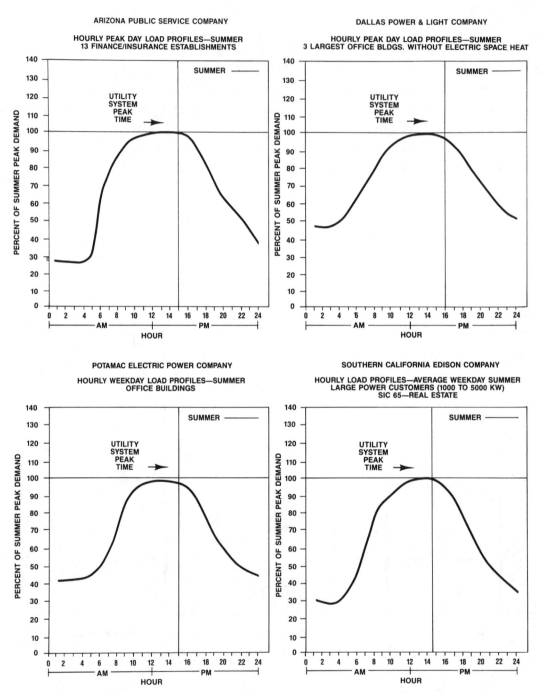

Figure 3-3.   Summer load profile for office buildings in four utility service areas

Buildings with 24-hour occupancy have cooling loads distributed throughout the day and night. Nonetheless, many of them have pronounced cooling load peaks during the utility's on-peak period, due to weather, and would be good candidates for cool storage, particularly because they tend to require smaller storage, minimizing capital costs.

Generally speaking, cool storage is not likely to be cost-effective in buildings with small cooling loads, in those with uniform daily load profiles, or in others (such as night clubs and theaters) where peak cooling loads occur during the off-peak hours. In addition, it will not be effective for buildings located where the summer season is short. Even if their peak cooling load profiles look promising, the short summer will reduce annual utility savings.

COLLECT BASIC INFORMATION

Assuming that threshold criteria are met, making a feasibility analysis worthwhile, basic information must be collected. Three of the most important factors to consider are: cooling and noncooling loads, economics (initial and operating costs), and local code compliance.

Another significant factor influencing design and specification is the system's adaptability to the conditions of an existing or proposed structure. These and other factors are listed in Table 3-1. Some of the major concerns are discussed below.

Identify Owner Preferences and Concerns. It is essential to learn of owners' preferences and concerns before a feasibility study is undertaken, to obtain information that will influence system selection and analytical methods.

Desire to Innovate. Some owners have a strong desire to innovate, and thus would be willing to rely on cool storage even though it is not as cost-effective as a conventional system. In such cases system design considerations will be affected by much more than financial considerations. The most cost-effective cool storage system may not be the one such an owner would prefer.

Aesthetic Impact. Any given cooling system -- conventional or otherwise -- can affect the exterior or interior appearance of a building. Because of the storage tanks, cool storage systems tend to have a more pronounced effect than others, and one that it is more difficult to conceal architecturally or support structurally. Is the owner willing to accept the aesthetic impact, or the cost of concealing it?

Table 3-1

General Information Requirements

- Owners preferences and concerns
- Building details
- Building usage and occupancy schedules
- Connected electrical loads
- Design day cooling load profile
- Design day noncooling load profile
- Local weather data
- Equipment performance data (full load and part load)
- Existing HVAC equipment
- Project constraints (space availability, architectural integration difficulties, etc.)
- Utility bills (if available)
- Equipment costs
- Electricity and Water Costs
- Heat recovery potential and use

Flexibility. How much flexibility does the owner want his system to provide? In many cases usages and occupancy patterns will change over time, and most owners will want to be able to accomodate such changes without undue expense. What does the owner have in mind? How will it affect system design? What about the potential for expansion?

Maintainability and Reliability. Although cool storage systems have been in use for many years, chances are the owner is unfamiliar with them and will regard them as something new. If he is greatly concerned about their reliability and maintainability, he may require a demonstration of one in use.

Comfort Requirements. Factors that generally have the most influence on comfort are:

- space temperature,

- space humidity,

- quantity and quality of temperature control,

- air motion control,

- mean radiant temperature,

- noise level,

- odor level, and

- air cleanliness.

For reasons of owner preference, unusual operating conditions, or others, will any of these factors take on more precedence than others? Will aspects of a cool storage system affect any in such a way as to be a concern? For example, maintenance of relative humidity as well as air supply and motion may be concerns if lower temperature air is provided in various spaces. Likewise, if an ammonia system is used, odor -- even potential toxicity -- may be a problem, requiring special isolation precautions.

Special Operating Conditions. The owner may have special operating conditions in mind and, if so, they should be ascertained. For example, will space cooling be provided only during certain months? Will no cooling be provided during evening hours? Will special operational requirements be instituted for use during other than normal business hours?

Economic Criteria. Numerous methods can be applied for purposes of cost analysis, and each is likely to render a different answer to the question, "Is it cost-effective?" The method which the owner is most comfortable with, and/or which he feels is most appropriate for the nature of the study involved, is the method to employ. Factors unique to an owner -- such as discount rate -- must be ascertained.

DEVELOP BUILDING LOAD PROFILES

Both cooling and noncooling building load profiles must be developed.

Cooling Load Profile

A knowledge of actual or projected annual peak cooling load (design day cooling load) and the maximum daily integrated cooling load is essential for proper comparative analysis of conventional and cool storage systems.

Peak cooling load is used for sizing a conventional system's refrigeration components. The number of ton hours of cooling effect which must be produced in a 24-hour period to meet maximum daily integrated cooling load is used for sizing cool storage equipment.

In all cases, the cool storage system must be capable of handling both maximum daily integrated demand and the design day cooling load, which may not occur on the same day. Both of these can be calculated using either manual techniques or computer programs. Design day cooling load can be calculated using procedures described in the ASHRAE Handbook, Fundamentals Volume, 1981 (8). To calculate maximum daily integrated cooling load, the designer needs to know the 24-hour load profile. This can be derived from detailed hourly calculations which many computer programs can perform.

It is important for cooling load calculations to account not only for heat gains to the space (solar, ventilation, lights, etc.), but also heat gain from fans and pumps. Any "safety factors" used in load calculations should be incorporated with care.

Calculating an excess cooling load will result in oversized equipment which, in turn, will lead to excessive fluid flows and wasted energy.

A typical design day building load profile for an office building (Figure 3-4) may be divided into four phases: cooldown, occupancy, shutdown, and off-time.

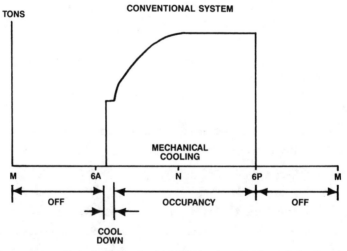

Figure 3-4.   Typical design day building load profile for an office building

Cooldown occurs prior to occupancy, when the cooling system brings a space to design temperature by removing accumulated heat. The cooldown period generally is kept as short as possible, with part or all of the system operating at actual full load for peak efficiency.

Just prior to the start of occupancy, the outdoor air dampers are opened and building lighting is turned on. The ventilation (or intake of outdoor air) represents an instantaneous cooling load that increases as the day progresses. The lighting and equipment heat gain which can be translated into instantaneous cooling load may never be 100% during occupied hours. Instead, the portion of lighting and equipment heat gain equal to instantaneous cooling load will vary from about 50% to 95% during occupied hours, depending on the mass of the building construction, the mass of furnishings and the space air circulation rate. In general, the cooling load trend between cooldown and shutdown will increase with time at a decreasing gradient until the time of peak load, and then may decrease slightly.

Shutdown usually occurs when the operating engineer decides that no additional cooling is needed, even though heat gains may still be present in the building. Shutdown causes the load profile to immediately go to zero.

Off-time occurs when the system is off, so there is no cooling load. Nonetheless, there always is heat gain, due to the release of lighting and solar heat gains previously absorbed by the building mass and furnishings. During the night, residual heat gain from lighting will average 10% - 20% of the daytime energy input. This becomes residual heat to be removed during the next day's cooldown cycle.

Noncooling Load Profile

Because an investment in cool storage is evaluated based on demand and energy savings, it is necessary to develop noncooling load data to identify the rate steps that will affect costs and savings.

Hourly noncooling loads can be estimated based on an analysis of energy-consuming equipment in the building and the extent to which loads will be used each hour of the day, each day of the year. The analysis is used both for determining hourly electric demand (for loads using electricity) and internal heat gains (from all loads). Although some noncooling loads, such as lighting, may be reasonably easy to estimate on an hourly and annual basis, other noncooling loads may present more difficulty. In such cases, it will be worthwhile to discuss certain types of equipment with building owners, managers or others who use it, to derive realistic measurements of its demand and performance.

The same computer programs available to calculate cooling loads generally can calculate noncooling loads as well. Many of them also have the capacity to perform 'energy consumption and cost data. These programs are discussed below.

SELECT STORAGE MEDIA

Chilled water and ice storage typically will both be candidates for cool storage projects. The choice will depend upon detailed study of site-specific factors. some of the factors involved are given below (5) (9).

Space Availability

Theoretically, one pound of ice is required to store 144 Btu of cooling. Practical considerations must be observed, however. For example, most ice builders are about half ice and half water when fully charged. Accordingly, the total mass required to store 144 Btu of cooling actually is about two pounds, not one. Likewise, when chilled water storage is employed, 8 pounds of usable water theoretically are required to store 144 Btu of cooling, assuming an 18F temperature differential. Due to mixing of return water with stored water, however, the usable volume of chilled water storage ranges from 70% and 85%, depending on design. Assuming an average volumetric efficiency of 80%, then, the actual storage required for 144 Btu of cooling would be 10 pounds of water, or about 5 times that of ice storage. Nonetheless, the actual size difference may be smaller than that implied by thermal considerations alone. For example, some ice storage systems are modular and require space for piping and maintenance. Still, in all cases, an ice storage system's storage requirements will be from one-fifth to one-eighth those of an equivalent-capacity chilled water storage system. Smaller storage requirements can comprise a distinct advantage, especially when retrofits are involved and underground storage is not feasible and above-ground space is at a premium.

Storage Stability

Storage losses must be estimated with care. With ice storage, dispersion of the ice volume within the water volume is such that thermal gradients and convective action are virtually nonexistent. By contrast, significant thermal gradients may exist in a chilled water storage tank, causing convective action to produce significant losses in a short period of time (Figure 3-5). Also, chilled water storage tanks are affected more by parasitic heat gain through tank walls, because the tank's surface area may be up to four or more times that of ice storage.

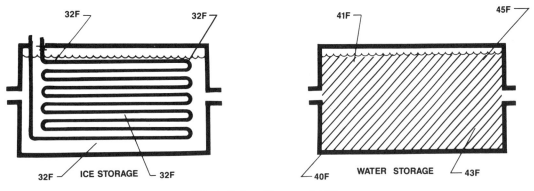

Figure 3-5.   Storage stability

Temperature Blending

Temperature blending is not a concern when ice storage is used. It is when chilled water is the storage medium. A number of techniques are available to minimize any potential problem. These are discussed in Section 5. Note that their cost and effectiveness vary. The dollars involved must be considered with care.

Efficiency

Water chillers use less kW/ton than ice storage systems due to higher evaporating temperatures. However, the evaporative condensers used with direct expansion ice storage systems have lower condenser temperatures and are more efficient than the cooling towers typically used for water chillers. These and other trade-offs must be considered in assessing annual energy efficiencies. Table 3-2 presents data on full-load power requirements of three alternate systems in an area with 1,000 equivalent fullload cooling hours (5). As can be seen, the primary chilled water pump's annual energy consumption is estimated to be (0.022 kW/ton x 300 tons x 1,000 hours =) 6,600 kWh. At $0.06/kWh, annual pumping cost will be $396. Annual energy costs for other components can be estimated similarly.

Chilled Water Temperature

Ice storage systems can supply lower water temperatures, and systems can be designed for larger supply/return temperature differentials (18 - 20F) that result in lower flow rates and pumping costs. Lower supply temperatures also allow for economical use of heat exchangers to offset high static head on tall buildings, and can effectively meet coil load requiring moisture removal.

Table 3-2

Full-Load Power Requirements (kW/ton) for
Alternate Storage Systems

	CHILLED WATER	ICE STORAGE SYSTEM	ICE STORAGE SYSTEM WITH HEAT EXCHANGER
CHILLER/COMPRESSOR	.83	1.3 (1.01)	1.3 (1.01)
PRIMARY CHILLED WATER PUMP	.022	—	.048
SECONDARY CHILLED WATER PUMP	.24	.24	.046
CONDENSER WATER PUMP	.038	—	—
COOLING TOWER FAN	.058	.044 (.044)	.044 (.044)
AGITATOR	—	.033	.033
TOTAL	1.188	1.617 (1.327)	1.471 (1.181)

ASSUMPTIONS:

1. 15-story office building.
2. Building peak cooling load: 750 tons.
3. Building cooling plant:
   a. Chilled water system with two 150 ton chillers rated at 45F chilled water and 95F condenser water temperatures.
   b. Ice storage system with two 150 ton compressors rated at 25F suction and 115F condenser temperatures.
4. Primary chilled water pump: 480 gpm, 40 ft. head.
5. Secondary chilled water pump: 1,200 gpm, 40 ft. frictional heat, 180 ft. static head.
6. Ice storage systems are direct expansion with freon. Flooded ammonia are shown in parenthesis.

Refrigeration Compressor Size

Direct expansion refrigeration compressors are size limited; multiple units on large plants may be uneconomical requiring the use of centrifugal or screw compressors.

Equipment Cost

Ice storage systems tend to be less expensive, because they do not require refrigerant-to-water heat exchangers, and have smaller storage requirements.

Maintenance Costs

Depending upon the system selected and the availability of skilled maintenance personnel, the maintenance of ice storage systems could be lower than equivalent water storage systems.

SELECT OPERATING MODE AND SIZE EQUIPMENT

An operating mode (full or partial storage) must be selected before a cool storage system is sized to analyze its performance. Very often several operating modes have to be analyzed before the optimum can be identified. Key factors to consider in selecting a cool storage operating mode are discharge start-time, discharge rates during different periods of the day, and maximum discharge rate. Charging factors also are important, and include the charging period, charging rate, and charging start-time.

Based on the operating mode used, cooling equipment capacity is found by dividing the maximum daily integrated cooling load by the number of full-load hours the chiller is allowed to operate in that day.

$$\frac{\text{Maximum Cooling Load}}{\text{Operating Time (hours)}} = \text{Required Compressor Capacity (tons)}$$

If the cooling equipment is to be operated for 14 hours at 100% capacity and 10 hours at 50% capacity on the design day, there are (1.0 x 14.0) + (0.5 x 10.0) = 19.0 full-load hours available.

To illustrate the storage-sizing procedure, consider a hypothetical building load profile with the maximum daily integrated cooling load as shown in Figure 3-6 (5). Operating mode calls for the chiller to be operated as described (14 hours at 100% capacity and ten hours at 50%). Superimposed on Figure 3-6 are the cooling equipment operation profiles and the annual peak cooling load. The area bounded by the cooling load and the cooling equipment output profiles represents the minimum required thermal storage capacity. This area approximates the maximum daily cooling load minus that portion supplied directly by the cooling equipment.

The maximum discharge rate is the maximum rate at which cooling can be supplied by the storage. As a minimum, this must be equal to the largest instantaneous difference between cooling equipment output and the peak cooling load. Thus, having determined the capacity and maximum discharge rate, the storage is defined from a thermodynamic standpoint. In an actual design, the storage size would have to be increased depending on the type of storage used, e.g., to account for the extra volume allowed for the empty tank in a tank farm, or the thermal stratification effects which occur in a chilled water tank (as discussed in Section 5).

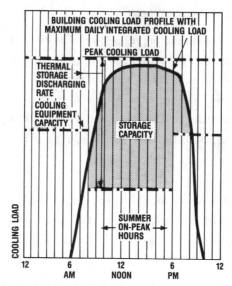

Figure 3-6. Thermal Storage and Discharging Rate

When determining the number of full-load hours that the cooling equipment is allowed to operate, any operating hours that cannot be utilized should be accounted for. As an example, Figure 3-7 illustrates a situation where cooling equipment is sized to operate 24 hours at full capacity just to meet the maximum daily integrated cooling load. Part of the operating mode prohibits storage charging during on-peak hours. As the cooling demand drops below cooling equipment capacity during on-peak hours, the cooling equipment will run at less than full capacity. This nonusable cooling energy, shown as a shaded area in Figure 3-7, must still be produced if the total daily cooling load is to be met. This requires the installed cooling equipment capacity to be increased so that these deficient ton hours of cooling are produced during the remaining available operating hours.

A more thorough discussion of operating modes and analytical methods involved in sizing is provided in Appendix A. Sections 4 and 5 provide specific discussion relative to ice storage and chilled water storage. Discussed below is an example illustrating the effect of different operating modes on equipment capacity and economics.

In this case, the example building is an office facility located in Los Angeles (5). It is studied using outputs from DOE-2.0A (a public domain computer program) to calculate monthly and annual electrical charges based on Southern California Edison Company time-of-use rate schedule TOU-8.

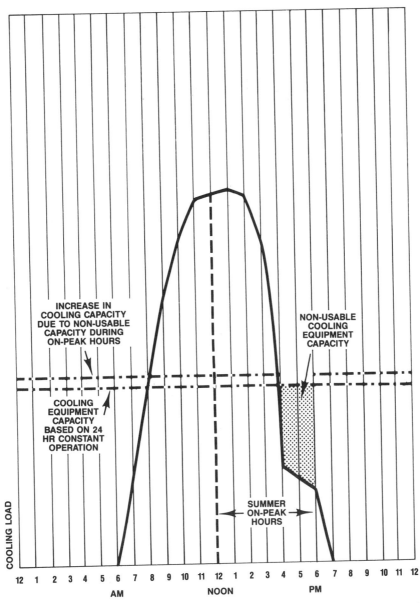

Figure 3-7. Increase in required cooling equipment due to partial non-use during on-peak hours

Capital cost differentials are estimated using $350 per ton of chiller and $0.60 per gallon of chilled water storage. Economic evaluations are based on simple payback.

The office building has 15 stories with a penthouse and two subterranean levels (Figure 3-8). Space conditioning is provided by variable air volume (VAV) systems

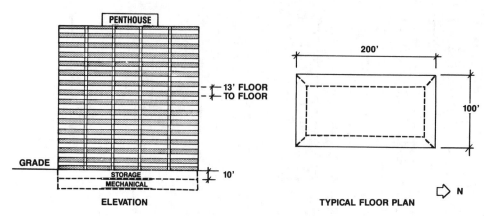

Figure 3-8.   15-story office building

with reheat (RH) in exterior zones. The central plant employs two electric drive
centrifugal chillers and two gas-fired boilers located in the basement. HVAC
systems and cooling tower are located in the penthouse. The general description of
the building architecture and energy systems are summarized as follows:

- Architecture: 15-story, steel frame, curtain wall, insulated
  concrete roof.

- Area: 300,000 square feet.

- Lighting: 3.0 watts per square foot.

- Equipment: 0.18 watts per square foot.

- System: VAVRH with economizer cycle, 78F cooling, 72F
  heating.

- Operation: 6 a.m. - 6 p.m.

Load profiles generated by DOE-2.0A are shown in Figures 3-9 (peak cooling load and
maximum daily cooling load) and Figure 3-10 (base noncooling load). As indicated,
the two cooling load profiles occur on different days; the peak cooling load is 8.7
x 10^6 Btu/hr, slightly higher than the peak of the maximum daily cooling load
profile. The integrated area under the maximum daily cooling load profile
indicates a maximum cooling load of 88 x 10^6 Btu/day. The building base noncooling
load profile shows peak electrical demand occuring at noon.

The cooling equipment capacity and storage capacity for three cool storage
operational modes (off-peak, mid-peak, and 24-hour operation), are established as
shown in Figures 3-11, 3-12 and 3-13. The performance of the storage systems for
different operating strategies were simulated, the reduction in demand and energy
charges determined, and payback periods calculated as shown in Table 3-3 (<u>5</u>). Note

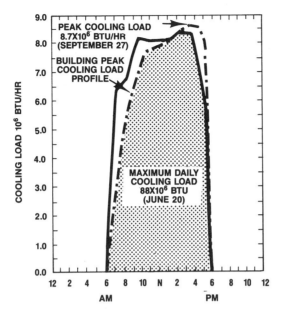

Figure 3-9.   Office building maximum daily cooling load & peak cooling load profiles

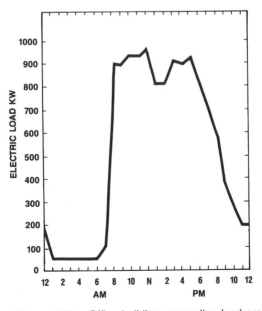

Figure 3-10.   Office building noncooling load profile

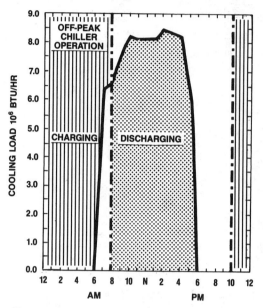

Figure 3-11.   Office building off-peak chiller operation

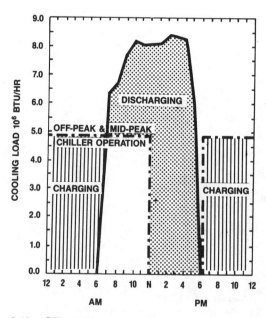

Figure 3-12.   Office building off-peak and mid-peak chiller operation

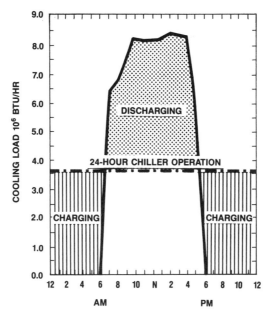

Figure 3-13. Office building 24 hour chiller operation

Table 3-3

Office Building Cooling Plant Comparisons

OPERATING STRATEGY	COOLING CAPACITY TONS	STORAGE CAPACITY TON HOURS	PAYBACK SYSTEM WITH YEARS
CONVENTIONAL	725	0	-
OFF-PEAK	750	7083	14.0
OFF & MID-PEAK	410	5833	8.3
24 HOURS	310	3917	4.1

that the cool storage systems are sized to meet the maximum daily integrated cooling load, whereas the conventional cooling system is sized to meet the annual peak cooling load.

Based on this preliminary analysis, it can be seen that 24-hour operation requires 43% of the conventional system chiller capacity, the storage capacity is 55% of that required for the off-peak system, and it has the shortest payback period. Once the most cost-effective operating strategy is determined, the other economic parameters can be included in a more rigorous analysis including life-cycle analysis to optimize the equipment and storage sizes. The present value of the life-cycle cost of the storage system with different combinations of cooling equipment and storage capacities is shown in Figure 3-14. The optimum size of cooling equipment and storage capacity is at the high point on the curve (310 ton capacity; 3,917 ton hours of storage).

With the thermal capacity of the storage system established, the next step is design of the physical system. The physical dimensions of the storage unit depends on the storage medium and range of operating temperature. If water is selected as the storage medium, the mixing of cold and warm water must be minimized.

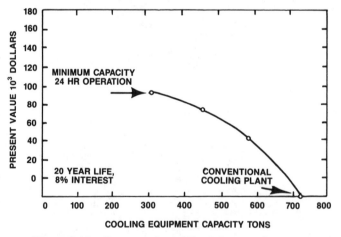

Figure 3-14. Present value of life-cycle cost of storage systems

CALCULATE ENERGY CONSUMPTION AND DEMAND

Energy calculations are performed in an iterative manner to examine not only various cool storage configurations and capacities, but also the impact that each of the various cool storage systems will have on building's air distribution system and on various rates available from the electric utility.

The basic approach to energy calculations is to establish demand and energy consumption on a monthly basis for the conventional cooling system (base case), and then do likewise for each alternative. Then, by applying appropriate utility rate structures, differences in energy costs become apparent.

A variety of manual and computer-based methods are available for performing energy calculations. Their accuracy, however, varies widely. It has generally been established that the accuracy of energy calculations for building need not (and most often cannot) be much greater than $\pm$ 15 - 20%. Nonetheless, it is particularly important to make accurate energy calculations for items which have substantial influence on cool storage system operating costs, electric demand in particular.

Where it is reasonable to assume that changes in the use of the building and/or changes in the electric utility rate structure may occur in the future, such changes should also be considered.

The manual methods used to identify energy consumption and cost are essentially the same as those employed to calculate cooling and noncooling loads. The procedures involved are detailed in the ASHRAE Handbook, Fundamentals Volume, 1981.

A number of energy analysis computer programs can evaluate cool storage systems. All can model the building and its cooling loads, as well as the types of conventional cooling systems and cool storage systems involved. Some of the functions performed by available public and private sector computer programs are shown in Table 3-4.

Since most energy analysis computer programs use hourly weather data, it is necessary to consider all factors which influence hourly cooling loads and hourly noncooling loads.

It usually is necessary to use the computer program on an iterative basis, with the first computer run generally being the conventional cooling system, and subsequent computer runs each consisting of a cool storage option. Results will indicate monthly electric demand and consumption and energy cost for each case considered. In addition, many computer programs can print out hourly summaries of the loads and energy consumption for as many days as desired. This makes it possible to examine conventional and cool storage system performance under a variety of loads and operating conditions.

Table 3-4

Functions Performed by Available Computer Programs

	PUBLIC PROGRAMS				PRIVATE PROGRAMS			
	DOE-2	BLAST	NECAP		AXCESS	TRACE	ESP-1	HCCJR	ESAS
HOURLY COOLING LOAD CALCULATION	X	X	X		X	X	X	X	X
SYSTEM AND EQUIPMENT SIMULATION	X	X	X		X	X	X		X
ELECTRIC DEMAND PROJECTION	X	X	X		X	X	X		X
GRAPHICS GENERATION	X	X			X				
ECONOMIC ANALYSIS	X	X	X		X	X	X		X

DOE-2:	U.S. Department of Energy, Lawrence Berkeley Laboratory
BLAST:	U.S. Army, Construction Engineering Research Laboratory
NECAP:	National Aeronautics and Space Administration
AXCESS:	Edison Electric Institute
TRACE:	The Trane Company
ESP-1:	Automated Procedures for Engineering Consultants, Inc.
HCCJR:	Automated Procedures for Engineering Consultants, Inc.
ESAS:	Ross F. Meriwether and Associates, Inc.

Whether energy calculations are performed manually or through computer, it is important to calculate energy consumption for both the cooling and noncooling uses.

Cooling Energy Consumption

In calculating cooling energy consumption, several factors need special consideration. One of these is the energy losses associated with cool storage, due to heat gains to the storage system. In addition, cool storage pumping energy consumption can exceed a conventional system's due to larger pumps, additional pumps, and/or longer hours of pump operation.

Cool storage systems' compressor energy consumption also often exceeds a conventional system's due to lower evaporator temperatures. However, cool storage systems' compressor performance can be improved when operated at night because outdoor dry bulb and/or wet bulb temperatures allow lower condensing temperatures. Examination of condenser fan and/or cooling tower fan energy consumption also is essential when comparing cool storage with conventional systems.

Any comparisons of conventional and cool storage systems also must account for part-load performance of cooling equipment. In conventional systems, cooling equipment operation follows the load. Continually varying loads thus cause less than optimal equipment operation, even with multiple cooling units. By contrast, a cool storage system operates at or close to full load during its charging period, resulting in close to optimal performance.

On most days full cool storage capability is not required. Cooling compressor load therefore can be based on maximum energy performance rather than either full load or just meeting the load.

Calculations also must consider that the compressor can be operated at higher suction termperatures (43-45F) in the partial storage mode, when the compressor is used for cooling water (as opposed to making ice or chilled water). Operation at higher suction temperature improves compressor efficiency. For example, a screw compressor operates 36% more efficiently when cooling water (at 43-45F) rather than making ice at 24F.

Experience with cool storage systems indicates they often consume more energy than conventional cooling systems. The cost of this additional energy usually is small, particularly when compared with demand savings. When time of day rates are involved, it is possible for the cool storage system's energy cost to be lower than that of a conventional system -- despite higher usage -- because the cool storage system's consumption occurs principally during off-peak periods, when rates are lower.

Note that, because a cool storage system can provide a colder cooling medium than a conventional cooling system, it may be possible to reduce fan and pump sizes, and their energy consumption, by using higher temperature differentials and reduced flows.

These smaller pumps and fans are often used to supply heating as well, thus permitting reductions in noncooling energy consumption year-round.

Noncooling Energy Consumption

Noncooling energy consumption can also be influenced by the choice of a cool storage system. The degree of detail required for such a calculation will depend upon the applicable electric rate structure.

When a declining block structure is used, so that unit rates decline as consumption increases, or where load factor-sensitive rates are used, so the cost of electricity declines when there are higher hours use of demand, it is necessary to estimate noncooling energy consumption more accurately, to enhance the accuracy of cooling energy costs.

Where time of day electric rates are used, energy consumption must be estimated by hour of the day and day of the week. Since some or all of the cooling will be accomplished during off peak hours, it is desirable to know how much noncooling energy is used during off peak hours so that energy costs can be properly calculated.

DETERMINE CAPITAL COSTS

In order to establish the economics of cool storage, it is necessary to estimate any changes in capital cost associated with cool storage compared with conventional cooling systems. In some instances capital costs will be lower and in others, higher. To the extent that there are increases in the capital cost for a cool storage system, the incremental capital cost should be used in the economic analysis.

The capital costs considered in cool storage include those associated with cooling equipment, storage and any additional space it occupies, piping and air distribution system, wiring and control, and design fees.

The cost of an ice storage system typically ranges from $1,100 to $1,400 per ton (10), with relative costs as follows:

- Major equipment   65%
- Materials   24%
- Labor   7%
- Miscellaneous   4%

Although incorporating cool storage adds new initial cost components to a building, in most cases it also permits significant capital cost savings, reducing the cool storage premium (if any) required. The savings involved are associated with the:

- electrical system,
- chilled water system,
- heat rejection system,
- structural system,
- air distribution system,

- architectural system, and

- acoustical control system.

These are discussed below, and Table 3-5 indicates the range of initial cost savings involved when compared to a conventional chilled water system (<u>10</u>). Note these cost savings are not cumulative since changes in one component can affect several others.

Table 3-5

Range of Cost Savings[1]

BUILDING SYSTEM	RANGE OF COST SAVINGS
ELECTRICAL	0 - 2%
CHILLED WATER	5 - 15%
HEAT REJECTION • PIPING • COOLING TOWER	 15 - 25% 20 - 35%
STRUCTURAL	0 - 1.75%
AIR DISTRIBUTION • COOLING COILS • FAN-MOTORS	 20 - 25% 4 - 9%
ARCHITECTURAL	0 - 0.75%
ACOUSTICAL CONTROL	0 - 0.3%

1. Expressed as a pecentage of the first cost of a conventional chilled water system

Electrical System: Because of load shifting with cool storage, less compressor horsepower is needed, meaning that smaller, less expensive transformers, switchgear, motor starters, and power distribution cabling and conduit can be used.

Chilled Water System: HVAC systems with cool storage can deliver colder supply water and use a larger supply-to-return water temperature differential. This permits reduced chilled water flow rates and, accordingly, smaller diameter pipe in the chilled water piping distribution system. For example, a conventional system with design supply and return temperatures of 42F and 56F, respectively, requires a flow rate of $\pm$ 1.7 GPM/ton. An HVAC system with ice storage can use design supply and return temperatures of 35F and 60F, respectively, requiring about 1.0 GPM/ton. Assuming equal water velocities for both systems, the 41.2% water-flow reduction would permit use of smaller diameter, less expensive pipe.

Heat Rejection System: When a partial storage operating mode is employed, the rate of heat rejection involved is far smaller than that associated with conventional cooling systems. This permits use of a smaller condenser water cooling system (cooling tower and piping), which results in cost savings. Certain structural system benefits also result.

Structural System: When conventional systems are employed, structural engineers typically must develop systems to support and often isolate 15 to 20 tons of rotating machinery. When cool storage is employed, far smaller structural loads typically are involved. For example, when a partial storage system is employed, the refrigeration compressors used (including their attendant motors and controls) commonly weigh 7,000 to 8,000 pounds, as opposed to the 25,000 to 30,000 pounds of the chillers they commonly replace. Likewise, the cooling tower used with cool storage usually is only half the size (sometimes even smaller) of the conventional-sized cooling towers otherwise required. The value of structural savings can be substantial.

Air Distribution System: Colder supply water temperature and larger water temperature differentials with cool storage permit first cost savings for both cooling coils and ductwork. Assuming the same criteria for cooling coil water tube velocities and air face velocities (i.e., same heat transfer coefficients), less coil surface (less dense fin spacing) will suffice, so the cooling coil needed is less expensive. Colder supply water temperature may permit cooler supply air temperature which, in turn, permits use of smaller, less expensive ductwork and air distribution systems.

Architectural System: In most instances the cool storage generation system is located on grade, in the least desirable space of a building. This approach may be able to free up valuable upperstory space, and make it even more attractive in that much of the conventional mechanical system noise problems would be eliminated. In addition, because of less mechanical system clutter on the roof, less screening is required to shield the equipment from view. In addition, when less equipment is exposed to weather, less weather-related maintenance is required, and longer life also may result.

Acoustical Control System: By locating compressors on grade, the extent of their noise and vibration transmitted through a building or at least to floors immediately below is significantly reduced, if not eliminated altogether. This enhances the attractiveness and value of upper-story space. At the same time, it reduces the design/installation/monitoring investment associated with direct and indirect noise and vibration control.

Ice Storage System Cost

Unlike chilled-water systems, whose size and design are unique to particular building applications, ice storage systems are available as "packaged" units from several domestic manufacturers. When installed, these units can be regarded as an integral equipment component in a total refrigeration-storage system. The modular, factory-built nature of ice storage systems, combined with their reduced space requirement, can represent distinct advantages over chilled-water systems for installation in existing buildings.

The unit cost of ice storage systems is steeply dependent on size, especially in the smaller capacity range. Thus, according to one manufacturer's price list, the unit cost of an ice storage system capable of holding 4,000 pounds of ice is $1.25 per pound of capacity, while the unit cost for a relatively large system, say 40,000 pounds, is only $0.45 per pound (6).

To obtain installed costs, the factory prices must be increased by the cost of transportation and installation. Transportation costs, which of course depend on distance, could add 3 to 10% to the cost of the delivered system. The cost of the installed system can then be estimated by assuming a 15% contractor's discount from the manufacturer's list price and then marking up the resulting price by 10% to cover installation costs, plus 10% for contractor overhead costs and 15% for contractor profit. Inclusion of these various cost components indicates that, for large systems, a representative cost value of $0.60 per pound, or $1200 per ton, of installed capacity, is appropriate (6). This corresponds to $50 per kilowatt-hour of electrical-equivalent storage capacity.

Chilled Water Storage Costs

The unit cost of chilled-water storage varies widely, depending on such factors as storage-tank material, storage size and location, and whether the storage is for old or new construction. Figure 3-15 presents representative installed-tank costs as a function of storage size. These unit costs exhibit significant returns to

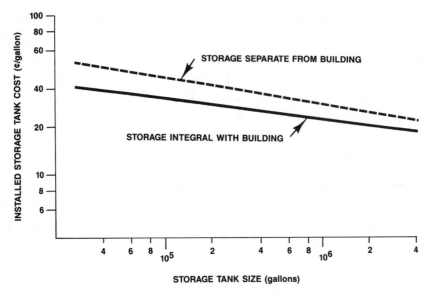

Figure 3-15.   Cost of chilled water tanks

scale over the range of capacity shown (**6**). Roughly speaking, 100 gallons of tank capacity are required to store the electric energy equivalent of one kilowatt-hour. Thus, a simple rule of thumb is that each 1.0 cent per gallon increase in the cost of storage corresponds to an increase of $1.00 per kilowatt-hour in the cost of storing an equivalent amount of electrical energy.

In addition to the "bare" tank costs, there are other storage-related costs, including the cost of the piping, pumps, and controls needed for operation of the storage tank. Again, there are large variations in these costs. A rule of thumb, based on experience with already-installed chilled-water systems, is that the cost of the other equipment components amount to 50% of the bare tank costs.

Refrigeration Equipment Cost

The storage systems must be matched with an appropriate refrigeration system and integrated into the complete building air conditioning system. Three types of refrigeration compressor systems -- reciprocating, centrifugal, and screw -- are currently available for use in commercial buildings.

In conventional air conditioning applications, centrifugal- and screw-type chillers exhibit roughly comparable cost and performance characteristics. Figure 3-16 shows representative installed-system costs of the three types of refrigeration

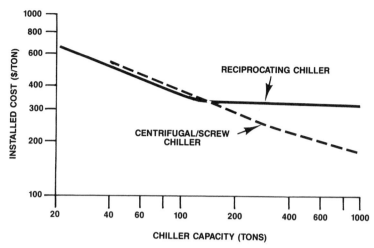

Figure 3-16. Cost of installed chiller systems

equipment. The system costs include the costs of the packaged chiller, the cooling tower, and the condenser water pumps, as well as installation costs, including contractor overhead and profit. As indicated in the figure, screw-type compressors are the lowest-cost alternative above 120 tons capacity (<u>6</u>).

Identify Costs and Savings Associated With Heat Recovery

Recovering waste heat from refrigeration equipment associated with cool storage can in many instances save energy and reduce operating costs. It must be remembered, however, that most or all of the cooling associated with cool storage is accomplished during off-peak hours. When load levelling cool storage is used, the recovered heat from the cooling plant can be used directly at that time if it is sufficient in capacity. If the quantity of recovered heat is not adequate during recover heat during off peak hours and possibly will require additional storage capacity for heating.

DETERMINE OPERATING AND MAINTENANCE COSTS

In order to accomplish an economic analysis it is also necessary to consider operating and maintenance costs. Those for cool storage may be higher than those for a conventional system, because more operating labor and supervision are required. Additional materials and supplies may also be necessary.

Identify Savings in Make-Up Water Requirements

Since cool storage equipment is usually smaller in capacity than conventional refrigeration equipment, its water requirements usually are less. The savings in cooling tower make-up water should be considered. The amount of make-up water required is based on the rate of evaporation and amount of bleed off. Water treatments required may also constitute an expense.

DOCUMENT VALUE OF ADDITIONAL BENEFITS

Although usually not deciding factors, there are a number of additional benefits of cool storage technology. The stored water in a chilled water storage system can serve as a standby water source for the fire protection sprinkler system, resulting in reduced insurance premiums. The stored chilled water can be used as a back-up source for cooling to critical spaces or equipment during power outages. Where possible, these benefits should be converted to dollar terms.

PERFORM ECONOMIC ANALYSIS

Many economic analysis tools can be used to evaluate cool storage systems. They range from simple payback analysis to life-cycle cost analysis. Note that many methods other than those discussed below may be preferred by the owner involved. These include cash flow analysis and discounted payback, as examples. Discussion of these and other techniques can be found in most handbooks of engineering economics. In some cases, however, owners will have developed their own methods which may incorporate elements of several "classic" techniques, for example, inclusion of tax benefits in formulation of a savings-to-investment ratio.

Simple Payback

Simple payback is the easiest of all economic tools to use and thus is widely used. It identifies the amount of time required for the savings generated by an option to equal the cost of implementing it. For analysis of a cool storage system it can be expressed as:

$$\text{Simple Payback} = \frac{\text{(Storage Cost - Ref. Equip. Cost Savings)}}{\text{Annual Savings in Electricity Bills}}$$

If the payback period calculated for a particular storage system is shorter than that normally required by the customer on similar projects, the customer may rationally decide to install a cool storage system. If the payback is too long, the customer may want to examine the feasibility of other storage systems, or may decide to drop the idea of cool storage altogether.

Despite its popularity, the payback criterion is not a formally rigorous basis for making investment decisions. It can, in fact, produce misleading results when used to compare mutually exclusive investment alternatives -- for example, the installation of either a load-levelling, demand-limited, or a full-storage cooling system. In addition, simple payback fails to consider the time value of money, the impact of inflation, or differential escalation (the difference between the general rate of inflation and the higher rate of inflation expected to affect energy prices in general).

Simple payback analysis therefore should be used only to obtain a very general idea about the cost-effectiveness of a given option. Other methods of analysis then are used for the more in-depth study that usually is required. The other methods used fall within the broad concept called life-cycle costing, or LCC.

Life-Cycle Costing

Life-cycle costing sums the energy costs of the building together with the net costs of purchase and installation (less any salvage value), maintenance, repair, replacement, and all other costs attributed to the investment. This includes the cost of money over the life of the investment. The investment that has the lowest total life-cycle cost while meeting the investor's objective and constraints is preferred.

All cash amounts are generally converted to net present value. Present value is defined as the equivalent value of past and future dollars corresponding to today's values. The net present value concept can be readily applied to evaluate the alternative cool-storage systems. If the discount rate is represented as R, then the net present value of an investment in a cool storage system with a lifetime of L years is given by:

$$\text{Net Present Value} = \sum_{N=1}^{L} \left(\frac{\text{Savings in } N^{th} \text{ Year}}{(1 + R)^N} \right) - \left(\text{Initial Investment Cost} \right)$$

In the analysis, the economic life and discount rate must be carefully selected.

Economic Life

The economic life is the length of time the equipment or device will last. Maximum economic life for a cool storage system is estimated to be twenty years, even though tanks will probably last indefinitely.

Discount Rate

Discount rate represents the rate of return one can expect to receive on the best alternative investment available.

"Discount rate" is an essential element of determining the present value of "future money." The value of money is time dependent for two reasons: first, inflation erodes the buying power of the dollar and second, money can be invested over time to yield a return over and above inflation. While it is possible that deflation might also occur, inflation is stressed because it expresses the more common condition in recent times. For these reasons, a dollar today will be worth more than a dollar received one year from now.

For example, if there were no inflation and investors could at best earn 10% interest per annum in a risk-free savings account, they would find a given dollar amount this year equivalent in value to that amount plus 10% a year hence. They could be expected to be indifferent between $100 now and $110 a year from now unless they had better investment opportunities available. A 10% rate of interest would indicate the investor's time preference for money. If there were additionally a 5% rate of inflation, investors would require $115 a year from now in order to be indifferent between that future amount and $100 today. The higher the time preference, the stronger the desire for money now rather than in the future and the higher the rate of interest required to increase future cash flows sufficiently to make them equal to a given value today. The rate of interest at which an investor feels adequately compensated for trading money now for money in the future is the appropriate rate to use for converting present sums to future equivalent sums and future sums to present equivalent sums, i.e., the rate for discounting cash flows for that particular investor. This rate is called the discount rate.

An illustration of the use of both economic tools for comparing alternative cool storage systems is provided in Appendix A. Because variations in building occupancy and rate schedules have tremendous impact on the economic viability of cool storage, a sensitivity analysis is also discussed.

Chapter 4

ICE STORAGE SYSTEM DESIGN

Ice storage systems -- as other cool storage technologies -- offer the potential of significant cost savings, particularly when lower-temperature water and air are utilized for space conditioning. If such benefits are to be obtained, however, a systematic approach is required. By assuring proper sizing of compressors, pumps, fans, pipes, ducts and so on, and by considering the interrelationships among these and other system components, optimized performance and operation will result.

Such systems comprise the most popular cool storage technology. Many system alternatives are available, but static ice storage systems are the most popular, with factory-built units usually being specified. However, when an installation requires three or more ice-builder-type factory-built units, cost comparisons should be made with field-fabricated systems. In either case, unit sizing and selection require knowledge of ice storage capacity data.

SIZING ICE STORAGE

As discussed in Section 3, storage capacity is influenced by building peak cooling load, maximum integrated daily cooling load and the operating mode selected. A typical ice-builder type unit using submerged pipe coils holds as much water as ice when fully charged. In a well-designed system, the average ice temperature typically is 6F below freezing, or 26F, for the conditions at full capacity. The absorption capacity per pound of ice is determined by adding 3 Btu/lb (6F x the specific heat of ice, 0.48 Btu/lb, F) to the heat of fusion of ice, which is 144 Btu/lb for a total of 147 Btu/lb of ice. Since tank volume is approximately 50% water and 50% ice, the water from the melting of ice plus the residual water in the tank warms from 32F to the highest useful level (after passing through the cooling coil), adding another 1 Btu/lb per degree F. The available heat absorption capacity per pound of ice storage based on highest acceptable chilled water temperature can be estimated from Table 4-1.

For the office building example discussed in Section 3, the size of the ice storage required for different operating modes can be determined as described below.

Table 4-1

Heat Absorption Capacity per Pound of Ice

FINAL TEMPERATURE °F	35	40	45	50	55
BTU/LB. ICE	153	163	173	183	193
LB. ICE/TON HOUR	78.4	73.6	69.4	65.6	62.2
TON HOURS/1000 LBS ICE	12.75	13.58	14.41	15.25	16.08

Using the load curve from Figure 3-9, the total ton hours required are first calculated as follows:

$$\frac{88 \times 10^6 \text{ Btu/hr}}{12,000 \text{ Btu/hr}} = 7333.3 \text{ ton hours}$$

The next step is determining the <u>refrigeration system capacity</u> for the three operating modes.

If a <u>full storage</u> operating mode is employed, total cooling load would be met without operating the refrigeration plant during the cooling period (6 AM to 6 PM). Refrigeration capacity would be 611.1 tons:

7333.3 ton hours ÷ 12 hours = 611.1 tons

When <u>load-levelling storage</u> is used, the refrigeration system can operate around the clock, resulting in refrigeration capacity of 305.6 tons:

7333.3 ton hours ÷ 24 hours = 305.6 tons.

For <u>demand-limited storage</u>, the refrigeration plant can run 18 hours (6 PM to noon) assuming summer peak demand occurs between noon and 6 PM. As such, refrigeration capacity would be:

7333.3 ton hours ÷ 18 hours = 407.4 tons.

Next, the <u>ice storage capacity</u> required for each of the three operating modes utilizing data from Table 4-1, is calculated as follows:

Full storage requires 7333.3 ton hours of ice. If 55F return water temperature is acceptable, then 16.08 ton hours is available from 1000 lbs of ice. The ice storage capacity required is:

    (7333.3 ton hours ÷ 16.08 ton hours/1,000 lbs) x (1,000 lbs)
    = 456,000 lbs ice

Partial storage (load-levelling) storage requires approximately 3500 ton hours which is the excess of the compressor capacity of 305.6 tons (3.66×10^6 Btu/h) under the maximum daily cooling load profile. The ice storage capacity is:

    (3917 ÷ 16.08) x (1000) = 243,594 lbs ice

Demand-limited storage requires approximately 4583.3 ton hours of cooling storage. This is calculated by estimating the percent of total cooling required between 12 noon and 6 PM, and the cooling required between 6 AM and that is in excess of the compressor capacity. The ice storage capacity required is:

    (4583.3 ÷ 16.08) x (1,000) = 285,032 lbs ice

A summary of the data for this example showing compressor size, storage capacity, peak load reduction for the three operating strategies is shown in Table 4-2.

Table 4-2

Summary of Example

	FULL STORAGE	LOAD-LEVELLING STORAGE	DEMAND LIMITED STORAGE	PEAK LOAD
COMPRESSOR OPERATION	12 hrs.	24 hrs.	18 hrs.	12 hrs.
COMPRESSOR TONS CAPACITY	611.1	305.6	407.4	725
ICE STORAGE	456,000	243,594	285,032	0
PEAK LOAD REDUCTION TONS	113.9	419.4	317.6	0
PEAK LOAD REDUCTION %	16	58	44	0
REDUCTION OF LOAD DURING ON-PEAK DEMAND TONS	725	419.4	725	0
REDUCTION OF LOAD DURING ON PEAK %	100	58	100	0

SELECTING ICE STORAGE EQUIPMENT

Several types of ice storage systems are available as discussed in Section 2. The
most popular as far as commercial buildings are concerned are the static ice
storage systems.

Four systems are finding increasing application in commercial buildings. One is a
static ice system which involves use of modular, insulated polyethylene tanks
containing a spiral-wound plastic tube heat exchanger surrounded with water (3).
It is available in three sizes -- 36, 54 and 100 ton hours. At night, a 75 percent
water-25 percent glycol solution from a standard packaged air conditioning chiller
which circulates through the heat exchanger and extracts heat until eventually all
the water in the tank is frozen solid. The ice is built uniformly throughout the
tank by the temperature-averaging effect of closely spaced counterflow heat
exchanger tubes. Water does not become surrounded by ice during the freezing
process and can move freely as ice forms, preventing stress or damage to the tank.
Typical flow diagrams for a partial storage system are shown in Figures 4-1 and 4-
2 (3).

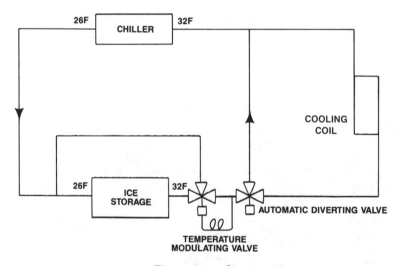

Figure 4-1.   Charge cycle

At night, the water-glycol solution circulates through the chiller and the heat
exchanger in the ice storage tank, bypassing the air handler coil. The fluid is at
26F and freezes the water surrounding the heat exchanger.

During the day, the solution is cooled by ice from 52F to 34F (Figure 4-2). A temperature modulating valve set at 44F in a bypass loop around the ice tank mixes with the 34F fluid, and achieves the desired 44F temperature. The 44F fluid enters the coil, where it cools air from 75F to 55F. The fluid leaves the coil at 60F, enters the chiller and is cooled to 52F.

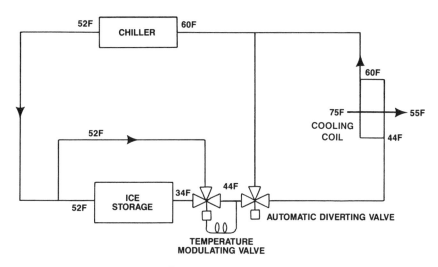

Figure 4-2. Discharge cycle

It should be noted that, while making ice at night, the chiller must cool the water-glycol solution to 26F, rather than produce 44 or 45F water temperatures required for conventional air conditioning systems. This has the effect of "derating" the nominal chiller capacity by approximately 30 percent. Compressor efficiency, however, is only slightly reduced because lower nighttime wet bulb temperatures result in cooler condenser water from the cooling tower resulting in higher unit operating efficiency. Similarly, air cooled chillers benefit from cooler condenser entering air dry bulb temperatures at night.

The temperature modulating valve in the bypass loop has the added advantage of providing unlimited capacity control. During many mild temperature days in the spring and fall, the chiller will be capable of providing all the necessary cooling for the building without assistance from stored cooling. When the building's actual cooling load is equal to or lower than the chiller capacity, all of the system coolant flows through the ice storage bypass loop, as demonstrated in Figure 4-3.

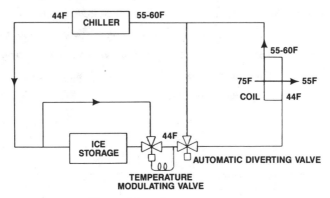

Figure 4-3.   Bypass loop

The glycol solution contains a multi-component corrosion inhibitor which is effective with most materials of construction, including aluminum, copper, solder and plastics. Unlike automotive-type anti-freeze, it produces no films and contains no anti-leak agents to interfere with heat transfer efficiency and permits use of standard system pumps, seals and air handler coils. However, because of the slight difference in heat transfer coefficient between water-glycol and plain water, chiller capacity should be derated by approximately 5 percent. It is also important that the water and glycol be throughly mixed before the solution is entered into the system. Additional discussion of this subject is included in ASHRAE Handbook, Systems Volume, 1984 (11) and in the ASHRAE Handbook of Fundamentals, 1981 (8).

The other three systems involve the use of prefabricated static ice-builder type units. They can be classified by water distribution and agitation strategy. These are single-compartment tanks using air pump agitation and a primary pump (Figure 4-4); two-compartment tanks using mechanical motor-driven agitators to control blend temperature plus primary pump (Figure 4-5); and multiple compartment tanks using the house pump for prime agitation and auxiliary low-heat pump for maintaining blend temperature (if necessary) (Figure 4-6).

All of these tanks are open with insulation on the bottom and sides, and with removable insulated covers.

Supplementing chillers typically are used with partial storage projects. The ice handles the demand-limited cooling requirements most of the time; the chiller is

used as an occasional off-peak supplement. However, since the chiller can use the
same compressor when generating ice, but has much better kWh/ton hour capability
when not generating ice, the use of supplemental chiller bundles help reduce
storage size. (See Figures 4-7 and 4-8.)

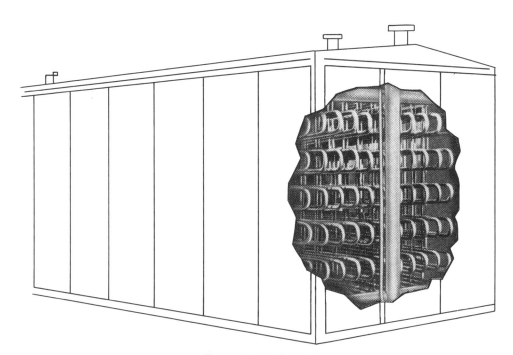

Figure 4-4.   Air agitated tank

Generally 1 1/4" pipe is used, but some models use 1" or 3/4". Circuit lengths
usually are in the range of 1,000 to 3,000 length/diameter ratios. The shorter
circuits can respond more rapidly to heavy loads, while longer circuits provide
more economical construction because of fewer connections.

Models usually are rated for 2" to 1 1/2" thickness of ice for storage capacity.
Note, however, that ice thickness affects pipe length requirements as shown in
Table 4-3. However, ice thickness should be governed by the ice building time as
shown in Table 4-4. A careful examination of the relationship of compressor size
to storage size should be undertaken. Operating at high suction temperatures is
desirable and should also be considered. The relationship of pipe size, ice
thickness and evaporator temperature is shown in Figure 4-9 for 12 hour recovery.
For other recovery times in hours (X), average evaporating temperatures T_2 can be

estimated as follows:

$$T_2 = 32 - (32 - T_1) \, 12/X$$

where T_2 = Average evaporating temperature for "x" hours
      T_1 = Average evaporating temperature for 12 hours

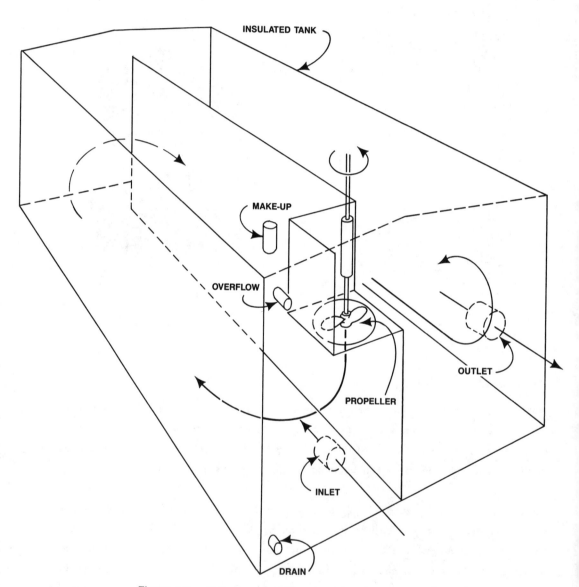

Figure 4-5.   Two-compartment tank with agitator

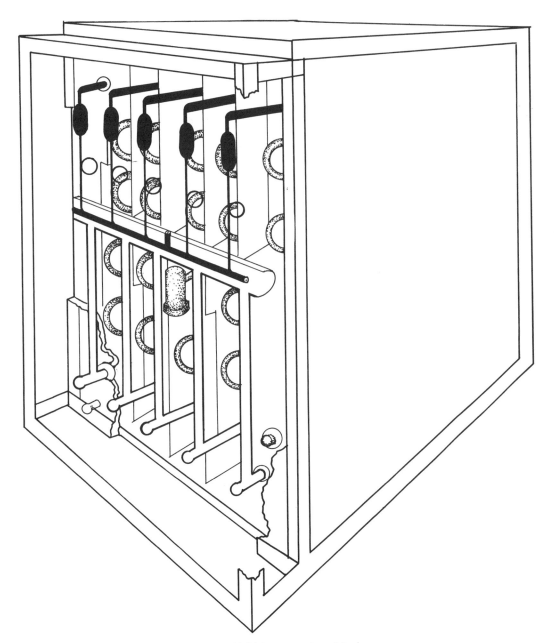

Figure 4-6.   Multiple compartment tank

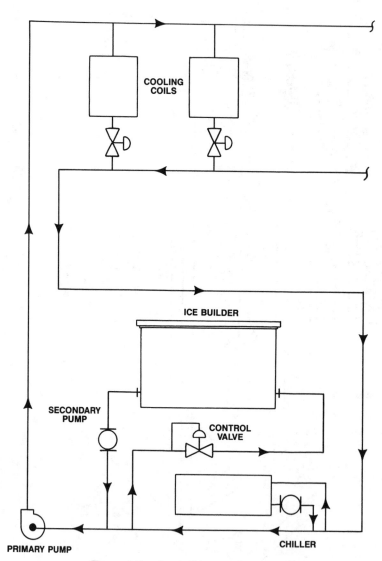

Figure 4-7.   Ice builder supplementing chiller

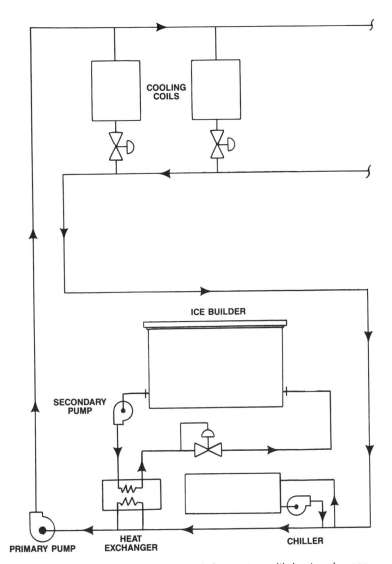

Figure 4-8. Twin loop circulating system with heat exchanger

Table 4-3

Pipe Length and Ice Holding Capacity

ICE THICKNESS INCHES	3/4" PIPE 1.05" O.D.		1 1/4" PIPE 1.66" O.D.	
	FT/1,000 LBS.	LBS./FT.	FT/1,000 LBS.	LBS./FT.
1	391	2.558	301	3.319
1 1/4	279	3.558	220	4.539
1 1/2	201	4.773	169	5.915
1 3/4	164	6.115	134	7.447
2	131	7.612	110	9.135
2 1/4	-	-	91	10.978
2 1/2	-	-	77	12.978

Table 4-4

Ice Building Time, In Hours, Using 1 1/4" Pipe Coils

ICE THICKNESS IN INCHES	EVAPORATOR TEMPERATURE °F			
	14	20	23	26
1.25	3.3	5.0	6.7	10.0
1.6	5.0	7.0	10.0	15.0
2.0	8.0	12.0	16.0	24.0
2.5	12.0	18.0	24.0	36.0

When return water is 42F or higher, auxiliary circulation from low head pumps or agitators is recommended to prevent "ice shock," a condition which causes ice to break up and float to the top, obstructing the water flow path.

In air agitated tanks, the return water should enter at one end and supply water exit from the other. When both water connections are at one end, a diverting channel is used to prevent short-cycling.

In two-compartment tanks, mechanical agitators usually pump 5 to 10 times the primary house flow, so the blend temperature is uniform throughout the tank. This contributes to even building and melting of ice.

Multiple compartment tanks, although simpler in design, generally do not yield their full rated storage unless the ice is regularly melted completely. This occurs because the ice remaining in the last compartment tends to get thicker, causing lower suction temperatures and increasing kW/ton.

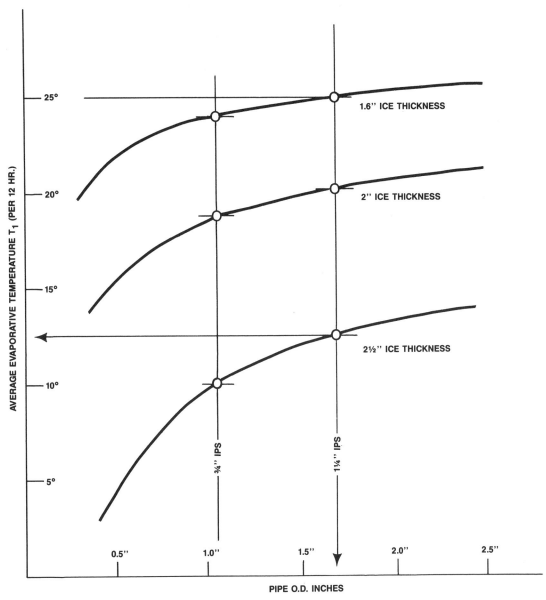

Figure 4-9.   Ice builder performance per pipe size for 12 hour recovery

The operating costs of agitation by air, low-head pumps, or propeller agitator are about the same. The choice therefore is based principally on system needs. Air agitation is easiest to apply to odd-shaped tanks. Mechanical agitation is superior if the tank is prefabricated (100,000 to 600,000 lbs.) and is shipped "knock-down." This is because of the simplicity of propeller agitators and their efficiency (5 hp can move 6,600 GPM).

Selection Factors

A number of factors must be evaluated in deciding which ice storage system to use. These factors include:

- Efficiency. Newer systems have features which make them more efficient than there predecessors, e.g., some of the newer units, because they rely on higher average evaporator refrigerant temperatures (24F vs 15F) use less energy during ice building.

- Modularity. Modular cool storage systems generally are more flexible because they can better match the building's cooling needs. They also contribute to lower cost operation.

- Controllability. The easier it is to operate a cool storage system through reliance on simple controls, the better the system will perform. Simple controls also help minimize operator problems which can cause undesirable downtime.

- Maintainability. Effective operation of the cool storage system depends upon maintenance. Major maintenance needs should usually be met through manufacturer-provided service agreement. The cost and availability of such an agreement is therefore a concern. Also, the system should be designed to provide easy access to in-house personnel for performance of routine maintenance functions.

- Manufacturer support. Manufacturers who provide effective support are essential. They must be relied on to provide accurate design information (the lowest suction temperature needed for ice generation, discharge rate, ice build rate, heat transfer area, the length of time the discharge temperature can be maintained, etc.) in a timely fashion, to provide operation and maintenance manuals, staff training, as well as references to others who use their systems.

BALANCING ICE BUILDERS & COMPRESSORS

The heat transfer associated with making and melting ice in ice builders traditionally has been by empirical methods. As an example, Figure 4-10 indicates the performance of ice builders with flooded or overfeed refrigerant control systems and 1 1/4" pipe coils. The curves shown represent the ice building time and evaporating temperature for a given capacity ratio (defined as compressor capacity in tons divided by 1,000 lbs. of ice storage).

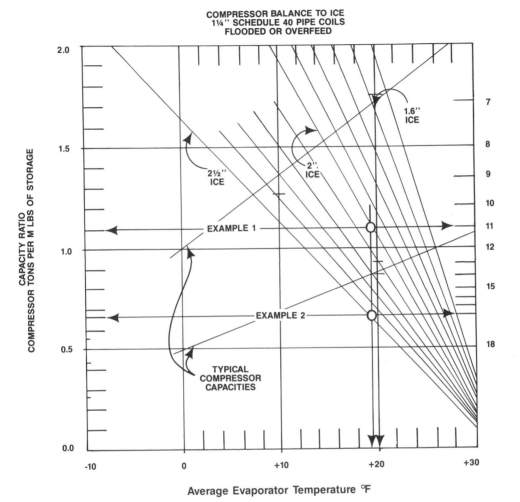

COMPRESSOR BALANCE TO ICE
1¼" SCHEDULE 40 PIPE COILS
FLOODED OR OVERFEED

(ALLOW 4° LOSS FOR R22 FROM EVAPORATOR TO COMPRESSOR FLANGE - ALLOW 2° LOSS FOR R717. DEDUCT 5° FOR Tx)

EXAMPLE 1   11 HOUR RECOVERY to 2" WITH 1.11 TONS PER 1000 LBS OF STORAGE WILL AVERAGE 19.6°F
EVAPORATOR TEMPERATURE.
• SELECT R22 COMPRESSOR FOR 15.6°
• SELECT R717 COMPRESSOR FOR 17.6°

EXAMPLE 2   18 HOUR RECOVERY TO 2½" WITH 0.67 TONS PER 1000 LBS OF STORAGE WILL AVERAGE 19.6°F
EVAPORATOR TEMPERATURE.
• SELECT R22 COMPRESSOR FOR 15.6°
• SELECT R717 COMPRESSOR FOR 17.6°

Figure 4-10.   Compressor balance to ice

The ice melting rate has seldom if ever been a determinant in selecting commercial building ice storage cooling systems. As a rule of thumb, about 60 - 75% of full storage can be melted in one hour, which generally can meet most peak requirements.

An example illustrating a method for selecting ice builder and compressors follows. In this example the 7,333 ton hour system shown in Table 4-2 is used. Further, the demand-limited storage mode requiring 285,032 lbs of ice storage, with an off-peak recovery time of 18 hours is employed.

Step 1: Determine capacity ratio

407.4 ÷ 285 = 1.43

Step 2: Determine evaporating temperature and ice-building time (from Figure 4-10)

1 1/2" ice requires 23F evaporator and 8 1/2 hours

2" ice requires 16F evaporator and 8 1/2 hours

2 1/2" ice requires 5F evaporator and 8 1/2 hours

Since R22 systems usually exhibit 3 - 5F loss in the return from the middle of the evaporator to the compressor suction flange, average suction temperature can be estimated as 4F below average evaporator temperature. Therefore:

1 1/2" ice: 23F - 4F = 19F suction temperature

2" ice: 16F - 4F = 12F suction temperature

2 1/2" ice: 5F - 4F = 1F suction temperature

Step 3: Calculate required length of 1 1/4" pipe from (from Table 4-3)

1 1/2" ice: 169 x 285 = 48,165

2" ice: 110 x 285 = 31,350

2 1/2" ice: 77 x 285 = 21,945

Step 4: Estimate compressor size

The CFM displacement of reciprocating and rotary screw compressors (for 95F condensing and 0 - 32F suction saturated temperatures) is estimated using a formula suggested by William A. Young (Crepaco, Inc.):

$$CFM = q/(a + bT) \qquad\qquad (1)$$

where CFM = ft^3/min displacement (or swept volume)

       q = BTU/hr

       a = 2300 for screw, 2100 for reciprocating

       b = 80

       T = suction temperature

Calculating the screw compressor required for 1 1/2" ice at 19F:

$$CFM = 407.4 \times 12,000/(2300 + 80 \times 19)$$
$$= 1279$$

For 2 ice at 12F: CFM = 1500

For 2 1/2" ice at 1F: CFM = 2054

Initial cost is minimized when the incremental cost of fabricated pipe coil equals the incremental cost of compressor displacement. At 1985 prices, this balance seems to occur at 20 to 40 feet of pipe per CFM of compressor displacement. The 2 1/2" ice will have unacceptable operating costs, and and may cost more initially because more than one screw compressor will be needed for 2,054 CFM. (1700 CFM screw compressors require 700 HP motors and and are usually the largest size used.)

Since a compressor size may not be available to match desired CFM, the suction temperature needed to balance a given compressor CFM is calculated as follows:

$$T = (q - a \times CFM)/b \times CFM \qquad\qquad (2)$$

(See equation 1 for definition of terms.)

Storage tank volume is not significantly altered when more pipe on closer centers must be accommodated, providing less ice thickness is involved. Tank volume of 27 - 35 ft^3 per 1,000 lbs of ice usually is adequate for 1 1/2" - 2 1/2" ice thicknesses.

In summary, then, the demand-limited storage could comprise three 95,000 lb. ice builders, each with about 11,000 feet of 1 1/4" pipe for 2" ice thickness. Either a single 1,500 CFM screw compressor or multiple screws totalling 1,500 CFM could be considered. The reciprocating compressor displacement needed is 1,643 CFM and could be provided by three large direct-drive multiple-cylinder compressors.

The largest preassembled ice-builders hold up to 100,000 pounds of ice (or 24,000 gallons of water) with pipe spacing and length suitable for 2" or 2 1/2" of ice. In making final selection for these preassembled units, compressors, and condensers, manufacturers' catalogs should be consulted. Units designed for 1 1/2" of ice must be custom-designed.

EQUIPMENT ARRANGEMENT AND LOCATION

As discussed above, ice building systems come in a variety of configurations. The largest factory built systems currently available have a capacity of 100,000 pounds of ice and up to 1600 ton hours of cool storage. It usually is more cost effective to use large ice storage units rather than smaller ones. Also, parallel piping is recommended. Although series piping produces longer contact time and colder water, static head or water level variations between the units can cause serious overflow problems.

Many of the hydronic system design considerations applicable to ice storage apply also to chilled water storage. These are discussed in Section 5. Note, however, in an ice storage system it is essential for the coil bank to be well secured to the tank. Otherwise, as ice forms on the coil, it could be dislodged, given the upward pressure exerted due to the buoyancy of ice. This buoyance is caused because ice is less dense than water, which expands by approximately 9% when freezing. For the same reason, water in a tank must be kept at a level which will accommodate the higher level that will be reached when freezing occurs. Also, since most ice storage units are open tanks, measures to prevent system static pressure problems must be considered.

One way to avoid system static pressure problems is to locate ice storage units at the top of the system either on the roof or in a top floor mechanical room. This approach requires the building structure to bear the weight of ice storage. Most of this weight is borne by the structural columns of the story immediately below. The effect on column sizing in succeeding floors diminishes rapidly.

If ice storage units are located at the bottom of the system, in the basement or on grade outside, three static pressure solutions generally are considered: pressure reducing valve (PRV), pump/turbine combination, or heat exchanger.

With a pressure reducing valve (PRV) on the inlet side of the ice storage unit, system static pressure is dissipated by the PRV, preventing the water level of the ice storage unit from rising. A water pump then is used to provide static head. The pump's total head capability must equal pipe friction loss plus the static head drop across the PRV. This technique affords maximum first-cost economy in the water piping distribution system, because it maximizes system water temperature differential. It is most applicable in low- to medium-height buildings where the pumping horsepower penalty would be negligible. (See Figures 4-11 and 4-12).

The <u>pump/turbine combination</u> is often used in high rise buildings. It consists of a pump unit and turbine unit mounted on opposite ends of a motor with a <u>double-extended shaft.</u> Note the shaft should not have any couplings because of the high static pressure involved. System static head consumed by the turbine unit is used to partially drive the pump unit which then replaces an equal amount of static head. This method also offers first-cost water piping economy.

Although the initial cost of a pump/turbine exceeds that of a standard pump, reuse of the static pressure reduces water pumping energy consumption (See Figure 4-13).

The third method comprises use of a <u>heat exchanger</u> as positive separation between system pressure and ice storage unit pressure. This approach is initially the most expensive of the three, due to the cost of the heat exchanger and because separate pumps must be provided for both both primary and secondary water circuits. Because a water temperature difference across the heat exchanger reduces allowable system temperature differential, large water quantities and larger system pipe sizes are required. Also, water pumping energy consumption is highest due to increased water flow and system pressure drop on both sides of the heat exchanger. Temperature approach requirement precludes extracting sensible cooling from the ice storage after the ice is melted. (See Figure 4-14).

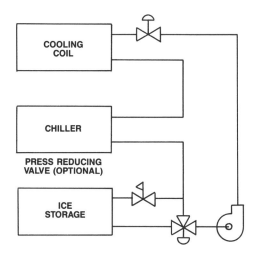

Figure 4-11. Direct water piping

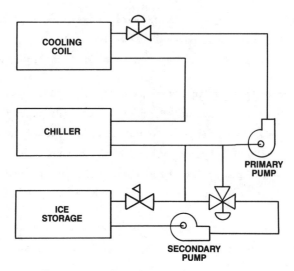

Figure 4-12.   Primary-secondary water piping

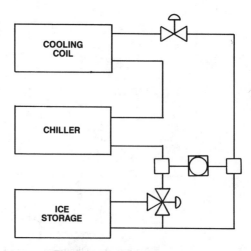

Figure 4-13.   Pump/turbine

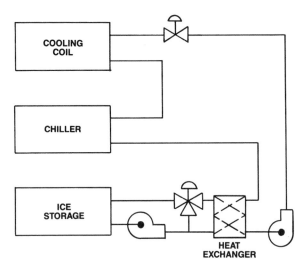

Figure 4-14. Indirect water piping

AIR DISTRIBUTION SYSTEM DESIGN

Ice storage systems can use colder supply water (or water-glycol solution) and --
when they do -- they also permit lower-temperature air to be distributed to
occupied spaces. Reliance on lower-temperature air permits reduced air quantities
which, in turn, can realize the significant first-cost and operating cost economies
associated with smaller air-handling units, ducts and associated equipment. Any
conventional air-handling system such as constant or variable volume water or air
can be used in conjunction with ice storage. The type of fan specified is based on
the type and size of air handling system selected. Smaller systems generally rely
on forward-curved blade fans, while larger systems employ backwardly-inclined or
more efficient air-foil-blade fans. In all cases, a variety of techniques are
available to help improve air-handling system efficiency, but a particularly sizable return on investment is realized when these are applied to systems using colder
air. Typical techniques include using straight duct runs, lower resistance coils
and high-efficiency filters to minimize resistance losses, reliance on high-quality
dampers and fittings, and application of duct sealers and proper levels of insulation to minimize air and thermal losses. In other words, by employing a systems
approach and applying effective -- but conventional -- air distribution system
design, substantial economies can be realized. (Refer to the <u>ASHRAE Handbook, Systems Volume,</u> 1984, Chapter 3, for details.)

It should be noted that use of lower temperature air is not without its drawbacks.
One of these is less air motion in the occupied space due to reduced air quantity.

This can result in complaints of "stuffiness" by the occupants. This problem can
be avoided through reliance on "terminal air blenders", which mix conditioned
air and return air in just the right proportion to meet the cooling requirements.

Sweating of ducts is another potential problem, when ducts are poorly insulated.
Larger latent load also is created by moisture migration into the building
and from outdoor air when room relative humidity is lowered.

The designer must perform a careful analysis of initial and operating costs and
comfort factors in deciding whether or not to use a lower-temperature air
distribution system. It likewise is important to carefully establish cooling coil
selection criteria and their impact on overall system economics.

The selection criteria that the designer must establish includes air side, liquid
side and capacity factors. Air side factors include coil air flow resistance, air
flow rate (cfm), face velocity, effective coil heat transfer area, and temperature
difference between entering and leaving air temperature. Liquid side factors
include liquid flow rate, entering and leaving liquid temperature, and liquid velocity
through tubes. Capacity factors include: total cooling capacity, sensible cooling
capacity, and cooling performance at various air and liquid flow rates.

Because of the diversity of factors that need to be evaluated in cooling coil
selection, it is best in most instances to avail oneself of the computer services
that major manufacturers of cooling coils provide. Some of the many factors that
should be taken into account in proper selection of cooling coils are:

    a.   The difference in temperature between the dew point of the entering
          air and leaving water temperature has a significant effect on coil
          performance. The performance improves when the coil is wet.

    b.   The configuration of the coil and fins, air velocity through the
          coil and the velocity of the fluid through the tubes determines the
          heat transfer rate. Obviously the greater the mean log temperature
          difference between the water temperature and air temperature, the
          greater the heat transfer rate. Water temperature closer to air
          temperature results in a lower heat transfer rate. Water
          temperature close to the dew point of air results in fin and tube
          surface temperatures being above the dew point of the air. Under
          this condition, the coil surface will remain dry with a resultant
          reduction in heat transfer rate. Until the air is cooled to its
          dew point, moisture removal cannot take place.

    c.   Reduced water flows reduce tube water velocity. A study of curves
          of coil manufacturers shows that the heat transfer rate reduces at
          an increasing rate as the water velocity reduces. One
          manufacturer's literature shows that a significant drop off for a
          coil begins to occur below approximately 3.5 fps. With variable
          air and water flow systems, the reduction in water flow could be

substantial. Variation in headering and insertion of turbulators inside the tubes can be used to minimize this problem.

d. Coils selected from tables require assumptions regarding tube velocity and water temperature differences. To validate these assumptions the designer must verify the tube velocity based on the water flow determined by the coil load and water temperature difference selected. When the actual velocities deviate from the assumptions, different velocities and/or water flow rates must be used if a suitable coil is to be selected. The velocity of fluid through the tubes determines the resistance of water through the coil and therefore the pumping head and ultimately the pumping cost.

The example given below illustrates the impact of different design strategies on unit and cooling coil selection, system pressure drop and fan horsepower. The building used for the example employs a variable air volume system. It is located in Washington, D.C., and is used principally for offices.

A summary of the building and system design data is shown in Table 4-5. The cooling load was calculated using manual procedures suggested by ASHRAE. Outdoor air quantity was calculated based on consideration of exhaust needs and local code requirements.

Table 4-5

Basic Building and System Data

BUILDINGS AREA OCCUPANCY	35000 FT^2 185 PEOPLE
DESIGN TEMPERATURE CONDITIONS OUTDOOR ROOM	 91F/74F 76F/64F
COOLING LOAD SENSIBLE TOTAL	 561330 597160
AIR QUALITY OUTSIDE AIR SUPPLY AIR	 5250 CFM[1] 31500 CFM
COIL AIR TEMERATURE ENTERING LEAVING	 78.5/65.6 57.0/56.6
WATER TEMPERATURE ENTERING	 48F

NOTE: 1. Based on toilet exhaust code requirements and exfiltration

Figures 4-15 and 4-16 illustrate outdoor and indoor design conditions and coil-entering and leaving conditions for two systems. One system employs conventional design using 48F supply water temperature and its air temperature leaving the coil is 65F. The other system uses 36F supply water temperature; its leaving air temperature is 47F. Several unit and coil selections were made to illustrate the effect of changing various design condition inputs. These are shown in Table 4-6, where conventional system unit and coil selection is shown in Column A. Column B illustrates a unit and coil selection identical to A, except the entering water temperature is 36F and a greater temperature differential (29F vs. 15F) is used. (capacity was held constant). This difference permits a reduction in water quantity and pressure drop, thus creating pumping horsepower savings (no air-side savings are possible). Column C indicates the effects of selecting a larger, more expensive unit with more coil face area and lower face velocity. As can be seen, air pressure drop through the system is reduced dramatically.

Column D is based on the same unit and coil selection but assumes a lower entering water temperature and a greater temperature difference. The resultant savings only relate to the water-side as compared to the selection in Column C.

Columns E and F illustrate the system design strategy employing 36F entering water temperature and a 27F temperature difference. Although internal loads and room temperature remain the same, a lower coil-leaving air temperature lowers the humidity ratio (dewpoint) of room air. Because outdoor air must be cooled to the dewpoint temperature of room air and since there is no change in internal cooling requirements, lowering the room air's dewpoint increases the total cooling requirement of the space. Accordingly, the only difference between Columns E and F is the unit size, coil face area, number of rows, and velocity. Nonetheless, a significant reduction in air side pressure drop occurs for the unit and coil selections indicated.

A comparison of Columns A and F clearly shows the tremendous advantages offered by using lower entering water temperature on both the air-side and water-side of the system, and providing lower temperature air in the space (47F vs. 56F).

Table 4-7 shows a computation of fan horsepower for three different unit and coil selections. Since the cooling coils are in the air flow circuit in each case, a lower air pressure drop results in considerable fan horsepower savings. As can be seen, the system supplying lower temperature air to a space (column F) also requires the smallest fan (17.5 brake horsepower) -- a reduction of 60% --

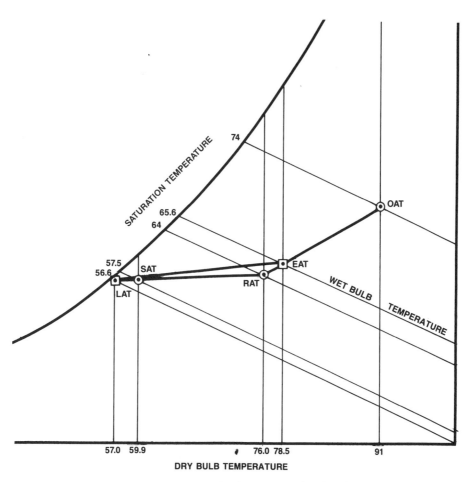

Figure 4-15. Psychrometric chart, conventional system
(31,500cfm, 48F entering water temperature)

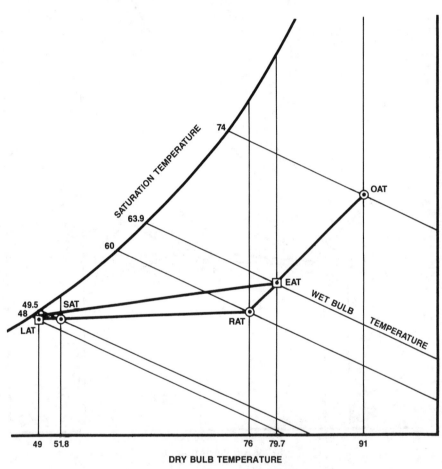

Figure 4-16. Psychrometric chart, alternative system
(21,087cfm, 36F entering water temperature)

Table 4-6

Air Handling Unit and Coil Selection

	A	B	C	D	E	F
UNIT SIZE	48 L	48 L	75 L	75 L	39 S	57 L
WIDTH	141	141	159	159	117	141
DT HEIGHT	81	81	104	104	71	81
COIL, SQ.FT.	56.9	56.9	90	90	39	61.4
FACE FT/MIN	554	554	350	350	541	343
COOLING COIL						
ROW/FIN	8 / 8	8 / 8	4 / 14	4 / 14	8 / 14	6 / 14
CIRCUIT	FL	FL	FL	FL	FL	FL
STD. AIR CFM	31500	31500	35100	31500	21087	21076
NON-STD CFM	31515	31515	31515	31515	21097	21087
ALTITUDE (FT)	14	14	14	14	14	14
CAP'Y BTUH	819600	891600	892570	891600	947010	947470
SENS. CAP'Y	750173	750173	750484	750865	739247	742107
U.S. GPM	115.97	61.11	122.24	63.00	70.59	68.69
PDW, WG.FT	17.39	5.78	6.66	2.12	8.38	5.79
VEL.FT/SEC	3.81	2.01	2.90	1.50	2.74	2.26
ENT. TEMP	48.00	36.00	48.00	36.00	36.00	36.00
LVG. TEMP	63.37	65.17	62.58	64.30	62.83	63.57
AIR ENT. DB	78.50	78.50	78.50	78.50	79.70	79.70
AIR ENT. WB	65.60	65.60	65.60	65.60	63.90	63.90
AIR ENT. H	30.35	30.35	30.35	30.35	29.05	29.05
AIR LVG. DB	56.85	56.85	56.83	56.83	47.83	47.69
AIR LVG. WB	56.46	56.46	56.46	56.46	47.82	47.61
AIR LVG. H	24.06	24.06	24.05	24.06	19.07	19.06
BYPASS FCTR	0.030	0.030	0.29	0.029	0.001	0.005
AIR FRIC. WET	1.00	1.00	0.34	0.34	1.31	0.49
AIR FRIC. @ ALT	1.00	1.00	0.34	0.34	1.31	0.49
AIR FRIC. DRY	0.81	0.81	0.26	0.26	1.06	0.37

Table 4-7

Computation of Fan Horsepower for Different Unit and Coil Selections

	A		D		F	
UNIT SIZE	48 L		75 L		57 L	
AIR QUANTITY(CFM)	31500		31500		21076	
AIR DIST. SYSTEM PRESSURE						
DROP (IN. W.G.)		2.50		2.50		2.50
COOLING COIL						
ROW/FINS	8 / 8		4 / 14		6 / 14	
AREA (SQ. FT.)	56.9		90		61	
FACE VEL. (FPM)	554		350		353	
RES. WET (IN.W.G.)		1.00		0.34		0.49
HEATING COIL						
ROW/FINS	2 / 8		2 / 8		2 / 8	
FACE VEL. (FPM)	554		350		353	
RES. DRY (IN.W.G.)		0.37		0.16		0.13
FILTER						
PRE-FILTER	66.6		105.5		66.6	
BAG FILTER 45% eff.						
FACE VEL. (FPM)	473.0		299		316	
BAG FILTER		0.44		0.39		0.39
PRE FILTER		0.30		0.27		0.28
TOTAL						
CASING (IN.W.G.)		0.20		0.20		0.20
FAN TOTAL STATIC (IN.W.G.)		4.81		3.86		3.99
FAN BHP[1]		43.5		34.0		17.5
EWT °F		48.0		36.0		36.0
GPM		115.97		63.0		68.07

NOTE: 1. Fan brake horsepower determined from carrier fan curves, with inlet dampers in wide open position.

resulting in less energy consumption and demand. (Fan selection was made using Figures 4-17 through 4-19.)

Because a variable air volume (VAV) system is employed, air delivery of the fan must be reduced to follow reductions of room loads. Since VAV boxes restrict air flow, fan air delivery is first reduced by increasing fan discharge pressure. The lower diagram of the fan curves shown in Figures 4-17, 4-18 and 4-19 indicate that the pressure/cfm point moves to a higher static pressure by following its rotational speed (rpm) line. A rise thus is required before the static pressure controller can cause the fan capacity reducing device to either start throttling the fan inlet vanes, reduce the fan speed, close discharge dampers or open a bypass damper depending on the method selected to reduce the fan capacity. The selection of a fan for VAV service requires a selection to the right of the fans peak static capability. If the fan is too large, it will not provide sufficient pressure rise (increase in differential pressure between the inlet and outlet of the fan). If the system static has been overestimated, the available pressure rise will be used up even before throttling is needed. Accordingly, all air handling units must be selected considering not only the coil size and capacity, but also the suitability of the unit fan for a VAV application. Selecting an oversized unit to attain low coil face velocity without consideration of fans available for use with that unit could result in an unsatisfactory fan selection.

The selection and sizing of other air distribution system components should also be based on careful evaluation of cost/benefits of the effects of reduced air volume and the lowered static pressure requirements of the fan. It is likely that the actual requirements for a reduced volume of air that is supplied at either lower temperatures or at lower static pressure will result in the use of forward curved blade fans instead of the use of fans with backwardly-inclined or air flow blades, most often employed in conventional air handling systems.

An indication of overall savings which can be obtained using lower temperature air (44F) rather than air at conventional temperature is shown in Table 4-8.

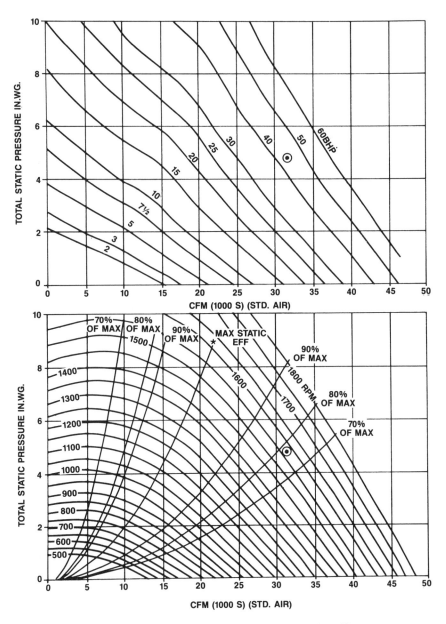

Figure 4-17. Fan performance, unit size 48L

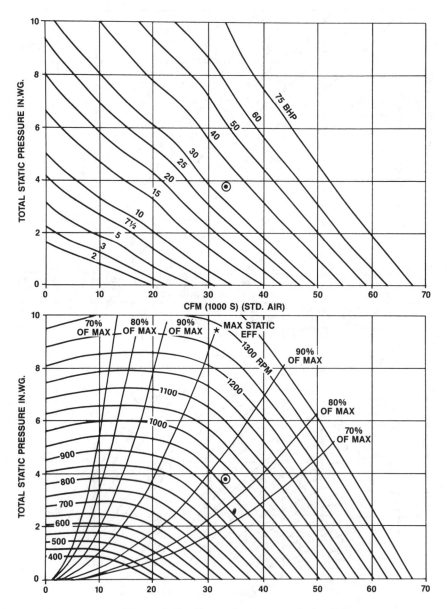

Figure 4-18. Fan performance, unit size 75L

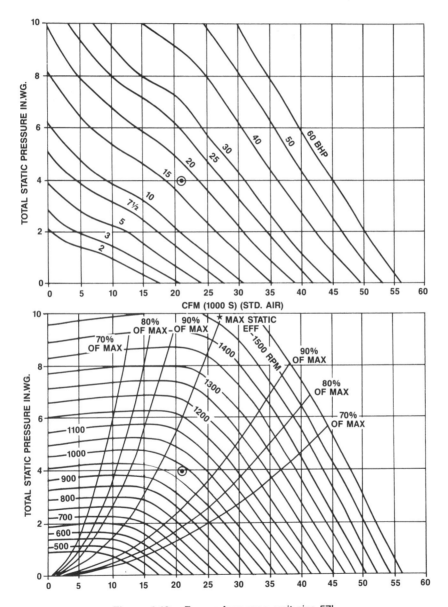

Figure 4-19. Fan performance, unit size 57L

Table 4-8

Summary of Comparative Analysis of a Conventional HVAC System
vs.
a Super-cool Air/Partial Ice Storage System Approach
for a 34 Story Office Building

GENERAL BUILDING DESCRIPTION: 34 STORIES ABOVE GROUND & 3 LEVELS UNDERGROUND PARKING.
TOTAL CONDITIONED SPACE   =   880,310 SQ. FT.
GROSS BUILDING SPACE   =   1,050,000 SQ. FT.
BUILDING PEAK A.C. LOAD   =   2,527 TONS

ITEM	CONVENTIONAL SYSTEM	PARTIAL ICE SYSTEM	COMMENTS
HVAC SYSTEM DESCRIPTION	MULTIPLE CENTRAL ALL AIR AHU's W/VAV DISTRIBUTION & CENTRAL CENTRIFUGAL RE-FRIGERATION CHILLER PLANT	CENTRAL SUPER-COOL AIR AHU's W/VAV & PARTIAL ICE STORAGE/SCREW COMPRESSOR REFRIGERATION PLANT	SUPER-COOL AIR APPROACH SAVED APPROX. 35,000 SQ. FT. OF RENTABLE SPACE
DESIGN DAY LOAD (24 HRS.)	30,185 TON HRS.	30,185 TON HRS.	
INSTALLED REFRIG. CAPACITY	2,600 TONS 2,006 PEAK KW	1,576 TONS @33F SUCT. 100° COND. 1,360 PEAK KW	40% LESS INSTALLED TONNAGE & 33% LESS PEAK KW W/ICE STORAGE SYSTEM
ICE STORAGE CAPACITY INSTALLED	-----	9,600 TON HRS. (680,000 LBS. ICE)	CLOSED SYSTEM W/48 ICE BUILDERS

INSTALLED DISTRIBUTION EQUIPMENT
a) PRIMARY CHILLED WTR. PUMPS
b) SECONDARY CHILLED WTR. PUMPS
c) CONDENSER WTR. PUMPS
d) COOLING TOWER FANS
e) EVAPORATIVE CON-DENSERS
    FAN MOTORS
    PUMP MOTORS
	
	97 KW
	
	116 KW
	
	168 KW
	
	87 KW
	
	----- 	
	----- 	
		
		61 KW
		
		76 KW
		
		----- 	
		
		----- 	
		
		70 KW
		12 KW	

A.H.U. SUPPLY & RETURN FANS	954 KW	437 KW	BASED ON 6.5" W.G. SUPPLY S.P. FOR CONV. SYST. & 4.5" W.G. SUP-PLY S.P. FOR ICE STOR. SYSTEM
FAN POWERED MIXING BOXES	70 KW	70 KW	
TOTAL DISTRIBUTION DISTRIBUTION KW	1,500 KW	727 KW	50% LESS INSTALLED DISTRIBU-TION KW W/ICE STOR. SYST.
ANNUAL OPERATING COSTS			
ELECT. DEMAND	5,246 KW	3,696 KW	30% LESS INCLUDES LIGHTS & EQUIPMENT
ANNUAL ENERGY(LIGHTS & MISC.)	9,486,940 KWH	9,486,940 KWH	
ANNUAL ENERGY (DIST. EQUIP.)	4,181,9833 KWH	2,062,289 KWH	50% LESS
ANNUAL ENERGY (REFRIG. EQUIP.)	6,515,052 KWH	4,690,788 KWH	29% LESS
ANNUAL ENERGY (HEATING)	442,647 KWH	442,647 KWH	
TOTAL ANNUAL ENERGY	20,626,582 KWH	16,682,674 KWH	20%
ANNUAL DEMAND COST	$364,812	$238,812	34% LESS
ANNUAL ENERGY	$1,938,362	1,556,936	20% LESS
TOTAL ANNUAL OPERATING COST	$2,303,174	$1,795,748	22% LESS W/ICE SYSTEM
ANNUAL SAVINGS USING PARTIAL ICE STORAGE a) DEMAND b) ENERGY c) SUBTOTAL (DEMAND & ENERGY) d) RENTAL SPACE SAVINGS	 ----- ----- ----- -----	 $126,000 $381,426 $507,426 $437,500	 DUE TO ADDITIONAL 35,000 SQ. FT. AVAILABLE RENTAL AREA
TOTAL ANNUAL SAVINGS	-----	$944,926	
COMPARATIVE BUILDING ENERGY USE	67,000 BTU/SQ. FT. PER YEAR	54,000 BTU/SQ. FT. PER YEAR	

WATER & BRINE DISTRIBUTION SYSTEMS

Use of lower temperature water distributed to air handling units permits larger water temperature differences. These in turn, result in lower water quantities and pumping horsepower. Pumping head is a function of system resistance to fluid flow through coils, pipes, fittings, etc., and the amount of resistance involved is significantly influenced by design. The pipe sizes are smaller with reduced flow when designed at conventional water velocities.

Note that closed water and brine systems require a means to accommodate the change in liquid volume which occurs as the temperature changes from a nonoperational to an operational status. Otherwise make-up water is required and air is introduced into the system. An expansion tank can be used for this purpose.

Further information on chilled water and brine systems is contained in Chapters 14, 17 and 26 of the ASHRAE Handbook, Systems Volume, 1984.

WATER PIPING CONSIDERATIONS

There are five basic water connections on an ice builder unit: chilled water outlet, return water inlet, make-up overflow, and drain. The chilled water supply and return water connection both require flow control valves to properly balance system water flow to and from the ice builder. The make-up connection is located above the water level and can be manually or automatically controlled. A valved drain connection is also required, to permit the ice builder to be emptied to a floor drain or other waste area. The ice builder is provided with an overflow connection which should be piped to a drain area.

Multiple Ice Builder Installations

There are two possible ways to connect multiple ice storage systems to the chilled water circuit: parallel and series. Parallel connection (Figure 4-20) is the least expensive approach to install, because it requires fewer valves and fittings. However, it is the most expensive method to operate, in that all ice builders melt down by the same amount when cooling is required. If the ice charge is only partially melted, then all ice builders must begin recharging with a partial thickness of ice on the coils. This partial thickness acts as an insulator in the heat transfer process and thus more energy per ton of ice is required during the re-charging process.

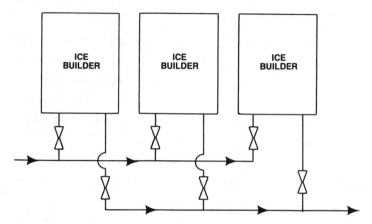

Figure 4-20. Multiple ice builder water piping, parallel arrangement

The series water flow arrangement and melt-down shown in Figure 4-21 is more efficient, but is higher in first cost. Each ice builder has isolation valves and a bypass valve to allow full water flow to pass through each ice builder individually and melt it down completely before the water is circulated through the next ice builder. This assures minimum power requirements during the rebuild process. However, no less than 25% of the total ice capacity should be isolated for meltdown at any one time, because any greater degree of isolation produces excessive water flow rates through the ice builder(s).

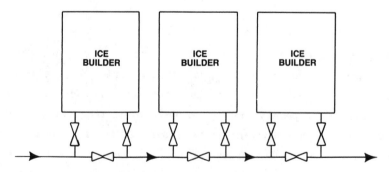

Figure 4-21. Multiple ice builder water piping, series arrangement

Equalization

When two or more ice builders are connected in parallel, a water equalizing line should be provided between adjacent ice builders as shown in Figure 4-22. This prevents the inadvertent ice builder emptying or filling which otherwise could result if flow control valves are not properly adjusted. It also prevents unnecessary overflow which could occur if the water flow rates from each ice builder became unbalanced due to an obstruction in the line.

The size of the equalizer connection is a function of the design water flow rate through one ice builder, and should be selected from Table 4-9. If ice builders are not the same size, equalizer connection size should be based on flow rate through the largest ice builder.

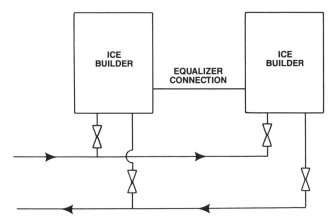

Figure 4-22. Multiple ice builder water piping, parallel operation, with equalizing connection

Table 4-9

Size Selection of Equivalizer Connection

ICE BUILDER FLOW RATE, GPM	EQUALIZER PIPE SIZE, IN.
Up to 120	3
121-240	4
241-630	6
631-1170	8
1171-1925	10
1926-2820	12

CONTROL

Conventional cooling system control strategies are selected to assure that chiller plant output meets cooling demand, and to help maximize the system's coefficient of performance (COP).

Several parameters must be controlled in order to maximize system COP. One involves control of chilled water temperature. This involves choosing the highest chilled water temperature that will satisfy the maximum load imposed on any cooling coil in the building.

Condenser water temperature is another consideration with selection generally comprising a trade-off between chiller unit(s) performance and energy requirements of a cooling tower fan. The lower the water temperature the higher the water chiller COP.

Selection and sequencing of chiller units in multiple chiller systems make up a third control consideration. On-line capacity should be matched to instantaneous cooling demand. In this way chiller units can be operated close to maximum system COP.

Control strategies developed for cool storage are significantly influenced by the physical composition and operational mode of the system involved, and by the building it serves. Factors common to all installations include storage system size, chiller efficiency, duration of cooling load period, and rate of thermal loss.

Adequate consideration and implementation of four separate control functions will help assure optimum performance of refrigeration and cool storage equipment. These control functions include:

- refrigeration equipment control during charging,

- next day building load prediction,

- optimum charge level calculation, and

- on-line load profile projection.

Refrigeration Equipment Control during Charging: This entails compressor/chiller control to maximize system COP. Controlled quantities are chilled water temperature, condenser water temperature, and unit selection/sequencing (when

multiple chiller plants are involved). The strategies employed are essentially
similar to those applied for conventional cooling, except cooling output can be
selected to <u>always</u> maximize COP (because chiller output does not have to match
building's cooling demand).

<u>Next-Day Building Load Prediction:</u> The rate of cool storage equipment thermal loss
affects overall system performance and efficiency. In turn, the rate of thermal
loss is influenced by the charge level; higher charge levels increase temperature
differences between storage media and their surroundings, increasing thermal
losses. Accordingly, it is desirable to limit charge levels. To do so, however,
operators must be able to accurately estimate cooling requirements (including
thermal losses) during the next period of peak cooling demand. This can be done
simply by employing "exponential smoothing." Successful use of this forecasting
technique assumes: (1)   that a building's cooling demand can be measured and
recorded as a function of time and outdoor temperature, and (2) that the data
recorded can be processed, as via computerized energy management and control (EMC)
system.

Expontial smoothing consists of three distinct steps:

1   Building cooling load and outdoor temperature data are recorded
     daily,

2   Cooling load data are least square fit to a load profile function,
     to produce coefficients which define the fitted load profile at
     different outdoor temperature conditions, and

3   Output from Steps 1 and 2 (comprising a sequence of coefficients
     for several days is statistically analyzed to develop a set of
     predicted coefficients for the next day.

In application, the process can be refined as more data are available for analysis.

<u>Optimum Charge Level Calculation:</u> Charge level is selected based on the predicted
next-day cooling requirement. The optimum charge level is that which delays the
charging process as much as possible (to minimize losses) while assuring cooling
storage will have the capacity needed to meet predicted requirements (including
consideration of the thermal losses which will occur after charging and before
use).

<u>On-Line Cooling Load Profile Projection:</u> Predicted next-day cooling requirements
will not always be correct, of course. When the stored charge is larger than
needed, no problems result. The load can be completely satisfied from the cool

storage system. When the charge is insufficient, however, the most appropriate alternative should be selected based on the shape of the cooling load profile. (Projecting the cooling load profile a few hours ahead of time can help assure the stored charge is used to obtain maximum economy, by minimizing peak chiller loads.)

Refrigerant Flow Control

When a volatile refrigerant system is required to operate at different suction temperatures and refrigerating capacities, the thermal expansion valve may operate erratically, causing more liquid to enter the heat exchanger than can be evaporated. This can lead to liquid passing through the suction piping and into the compressor, resulting in "slugging" of the compressor, a condition evidenced by a characteristic noise (in a reciprocating compressor). Use of an evaporator pressure regulator can stabilize operation and prevent slugging. This device is basically a valve which is modulated by suction pressure directly or through a pilot operator which can also be operated by temperature controllers in an air stream. Sizing of the valve is critical to proper operation. Excessively large valves cause "hunting"; small valves create excessive resistance to refrigerant flow and therefore lower compressor efficiency. Evaporator pressure regulators can also be used to reduce heat transfer of a device by raising the evaporator temperature (pressure), and they are particularly useful when two evaporators on the same piping system must operate at different suction temperatures, as when ice is being made and air is being cooled simultaneously. It may also be used to retain an evaporator temperature while allowing the compressor(s) to unload by lowering the suction pressure, thus allowing cylinders to unload and also stop compressors after being fully unloaded.

System Control

The manner in which a cool storage system is controlled depends upon the system itself, the mode of operation, and many other factors unique to the system and building involved. The following discussion relates to two of the most commonly used static systems -- ice builder and modular brine -- and is somewhat general in nature, principally to indicate considerations which should be applied. In all cases, the equipment manufacturers should be consulted to learn of the precise operating characteristics of the equipment.

Ice Builder System: The simplest ice builder system generates ice and then circulates chilled water through an air cooling coil in an air handling unit. A more complicated system cools air by utilizing a direct expansion coil in the air handling unit upstream of the water cooling coil. Some systems also permit cooling with both chilled water and volatile refrigerant at the same time. If this dual function system is used, the ice builder and direct expansion air cooling coil liquid pipes require a <u>refrigerant solenoid valve</u>. When the system permits simultaneous operation of both the ice maker and direct expansion refrigerant air cooling coil, an evaporator pressure regulating valve must be installed in the suction pipe from the direct expansion coil.

Most ice builder systems also require a manual selection switch for mode selection, <u>i.e.</u>: off, ice building, air cooling with chilled water, air cooling with direct expansion, and air cooling with chilled water and direct expansion.

In the <u>ice building mode</u>, the refrigerating cycle is activated by energizing the compressor and heat rejection cycle, and by opening the liquid refrigerant solenoid valve in the liquid line to the ice builder. With the chilled water pump off, the compressor suction pressure controller is set at the lowest pressure necessary to generate ice. The compressor operates until the ice thickness gauge (or whatever means has been selected for stopping ice making) closes the liquid line solenoid valve. This causes suction pressure to reduce, thereby unloading the compressor and then stopping it after the suction system piping is pumped down. The compressor remains off until again activated for ice making or air cooling. (Consult the compressor manufacturers for recommendations on compressor operation for this application.)

If the system must be operated at low ambient air temperature, a controller should be installed in the high-pressure refrigerant side to operate the mechanism chosen to reduce the heat extraction capability of the condenser or cooling tower. This will maintain that pressure differential across the thermal expansion valve needed to permit the thermal expansion valve to pass the required quantity of refrigerant.

   <u>Air Cooling With Chilled Water Mode</u>: In this mode, the air handling unit
   unit fan is in operation and automatic air temperature controls are activated.
   The chilled water pump is operational and the mixed water to the chilled water
   coil modulates the three-way (or two-way) valve to maintain the water at the
   temperature used to size the air cooling coil. The compressor is deactivated.

   With systems having multiple air handling units, a mixed water or supply air
   temperature controller operates a three-way valve to maintain the water at the
   temperature used to size the air cooling coils. The compressor is
   deactivated.

When an air handling unit is in operation, a room thermostat or supply air temperature controller operates a three-way or two-way valve near the coil to regulate the quantity of water through the coil. If the system has only one air handling unit, the controller can reset the control point of the mixed water three-way valve to regulate the supply air temperature.

Air Cooling With Direct Expansion Coil: In this mode, the air handling unit fan is in operation and air temperature controls are activated. The compressor suction pressure control is positioned to the pressure controller which is suitable for the direct expansion air cooling coil. Then the temperature controls are activated, the supply air temperature controller or the room thermostat opens the liquid line solenoid valve to the direct expansion coil and also energizes the compressor controls. Because the refrigerant suction pressure will be high, the compressor will start. Once the cooling load is reduced, the refrigerant suction pressure falls to a predetermined level thereby unloading the compressor. Upon a continued fall with the compressor pumping down the refrigerant system, a pressure controller stops the compressor. When cooling is again needed, the cycle repeats.

Air Cooling With Chilled Water and Direct Expansion: This cycle may be used to enhance system efficiency by operating the compressor at a higher COP during normal occupancy periods. In this mode, the air handling unit fan is in operation and air temperature controls are activated. The control sequence of both "Air Cooling with Chilled Water Mode" and "Air Cooling with Direct Expansion Coil" are activated, except the air-side controller first activates the direct expansion cycle and then the chilled water cycle. If cooling can be accomplished with the direct expansion cycle, the chilled water cycle will not need to operate. However, the compressor unloading pressure control point would need to be elevated to achieve a higher COP at other than full load operation of the compressor.

Modular Brine System: The modular brine system is discussed and illustrated in Section 2; its cycle is discussed earlier in this section. During the charging cycle, the leaving brine temperature controller should be set at the desired temperature (approximately 26F). Flow switches and a low-limit temperature controller are needed to shut down the refrigerating compressor when water flow is below that recommended by the chiller manufacturer or when the brine temperature approaches 16F. (Under low-flow conditions, a freeze-up can occur in the water cooler tubes.)

An <u>integral control panel</u> is customarily furnished with liquid chilling packages to regulate compressor capacity by first unloading and then stopping the chiller compressor. The control panel also interlocks <u>flow and temperature safety controls,</u> chilled brine and condenser water pump operation and cooling tower fans, as well as other safety features, such as lubricating oil pressure, etc., deemed necessary by the chiller manufacturer.

As charging occurs, the return water temperature begins to fall, causing the supply brine temperature to fall as well. This results in a controller unloading the compressor. When the compressor capacity reaches its minimum operating capability, it shuts down. With centrifugal chillers, this value is approximately 10% of full load at the brine and condenser water rating conditions for the chiller used. One method of stopping the chiller relies on a temperature controller which operates when the temperature between the supply and return brine reaches a point where further cooling would be uneconomical, as when all water in the storage tank(s) is frozen.

During certain seasons of the year it may be desirable to limit the production of ice to that needed for predetermined periods.

In such cases a Btu meter could be installed to measure the amount of heat extracted from the water, and to stop the chiller when no additional storage is needed.

The unloading capacity of reciprocating chillers depends upon cylinder availability for unloading as well as the quantity of compressors in the package. When multiple tanks are used, all are charged simultaneously. The brine flow rate should be held at the rate specified by the manufacturer of the ice storage system, because it is crucial for the ice manufacturing rate in all tanks to be held constant. In this way all tanks will contain the prescribed weight of ice at the end of the charging period.

If a full storage operating mode is used, the valving can be arranged to bypass the chiller and pump brine to air-handling units upon demand. In such cases a temperature controller is needed in the mixed water supply pipe to the air handling units. A three-way valve also is needed to bypass brine around the ice tanks and

to mix brine from the storage tanks(s) and return water from the system to provide
the desired brine temperature to the air handling units.

If a load-levelling or demand-limiting partial storage operating mode system is
involved, the chiller needs to operate while stored energy is also being used. In
this mode, the control point of the temperature controller in the chiller outlet is
reset to a temperature approximately 2F below that established as the design
temperature used for selecting the air cooling coils. As long as the chiller can
produce that temperature brine, the three-way valve controlling the flow through
the ice tank coil(s) will cause all brine to bypass the ice tank(s). When the
return brine temperature rises enough to cause the chiller outlet brine temperature
to rise above that established for mixed brine, the controller will begin to allow
brine flow through the ice tank(s). This arrangement permits the stored brine to
accept the peak load.

Control Devices

Many cool storage control devices are available. Selection is based on operating
mode considerations as well as initial and long-term cost factors.

Time-based controls typically are applied to start and stop equipment at
predetermined intervals. They also are applied to control the three-way valves
which direct fluid flows from storage to the compressor/chiller or air-handling
system cooling coils.

Microprocessor-based controls can perform all functions needed by time-based
controls, as well as the four control functions discussed above. They also can be
applied for optimizing functions, i.e., those which improve equipment or system
operating efficiency by monitoring performance and making adjustments on a real-
time basis. Most are capable of handling several other functions as well, such as
lighting control, duty-cycling, and demand control. Most microprocessor-based
controls are relatively low in cost.

Centralized computer-based controls can be easily programmed to handle cool storage
requirements, and can help optimize system performance by providing additional
functions such as monitoring, hard-copy logging, alarm, maintenance scheduling and
remote trouble-shooting. Distributed control systems usually are employed for
larger installations. These employ microprocessor-equipped field interface devices
(FIDs) to perform many logic functions on a by-exception basis, without reliance on

the system's central processing unit (CPU). This approach reduces overall burden
on the system CPU while also affording excellent system redundancy. <u>Direct digital
control</u> takes the distributed concept a step further. Microcomputers are installed
in local control loops to handle even more functions without reliance on the system
CPU. This approach also serves to enhance control precision. With or without
distributed or direct digital control, a centralized computer-based approach
affords the additional benefit of easy operating strategy modification, to obtain
optimum performance based on experience.

<u>Ice Thickness Control</u>: No matter what type of overall control strategy is used,
the ability to monitor and control thickness of ice build-up is a critically
important issue. A variety of measurement methods and controls are used in this
regard. These include:

> <u>Electric Resistance:</u> Standard ice thickness controls which rely on the
> difference in conductivity between ice and water are available. Probes
> are mounted at representative thicknesses, for example 1/2", 1", 1 1/2",
> 2", 2 1/2", 3". When ice grows to the selected probe, a circuit is
> completed and a control network signals that the desired thickness is
> attained. This ice thickness control can be wired to poll the ice
> builder and give an indication as to which thickness has been reached.
> A primary use of the thickness sensor (Figure 4-23) is to deactivate
> the compressor when the desired ice thickness is attained.

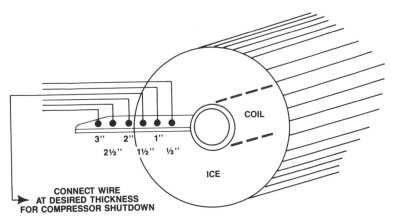

Figure 4-23. Ice thickness sensor

<u>Liquid Displacement:</u> Freezing water expands approximately 9 percent. One cubic foot of water will occupy approximately 1.09 cubic feet as ice. The increase in volume is a measure of how much the water in an ice builder has been converted to ice. The liquid level in a tank will rise as ice is formed. If the tank is full to the overflow line with no ice, the volume of water displaced divided by approximately 1.09 will give the volume of ice in storage. On melting, a system to replace the displaced water must be devised.

<u>Remote Bulb Thermostat:</u> The bulb is mounted usually three pipes down from the top and about 1/4" to 1/2" inside the desired ice thickness.

<u>Other Ice Thickness Measurement Methods:</u> Some designs use mechanical fingers with calibration dials to allow a manual check of the ice inventory. Other designs use transparent windows to allow visual determination of the ice thickness.

INSTALLATION REQUIREMENTS

The space required by ice storage tank must include an area around its perimeter to provide access for servicing, and -- in the case of multiple ice builders -- to provide a space for installing water equalizing connections. An access lane of at least three feet is recommended along each side and between adjacent ice builders (<u>2</u>). A minimum space of six feet in width should be provide on any size where the water connections are made. See Figure 4-24 for a typical layout.

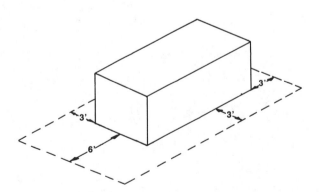

Figure 4-24. Minimum access requirements

Chapter 5

CHILLED WATER STORAGE SYSTEMS DESIGN

Chilled water storage systems have been used successfully for many years in numerous applications. Some of their many advantages when compared to alternative space cooling systems include:

- Conventional chillers, piping and air handling units can be employed. This permits better equipment selection and availability, which may result in more competitive pricing.

- Controls are simpler than those associated with an ice storage system. This helps assure that design demand reductions are attained, without sacrificing comfort.

- Reliance on conventional system components and controls makes it easier to find operating and maintenance personnel, and reduces the amount and complexity of training they require.

When designing a chilled water cool storage system, a number of factors must be considered. Blending, stratification and storage efficiency are particularly important concerns, and are discussed separately, below.

BLENDING

When water is the storage medium, blending of the warm water returning from the cooling coils with stored chilled water and dead spaces must be minimized. Several methods have been employed to minimize the blending problem (5) (12) (13), as follows.

Diaphragm-Type Systems

As the name implies, diaphragm-type systems use a diaphragm, usually made of a rubberized fabric, to physically separate warm return water from cool water. Vertical diaphragms that move back and forth inside a tank (Figure 5-1) are not generally recommended. Experience indicates that the diaphragm rubs against the tank walls and quickly wears through, and/or gets caught in inlet pipes and ruptures. The latest designs use horizontal diaphragms that move up and down in the tank, as shown in Figure 5-2. Since the warm water is stored above the cool water, this approach tends to keep the two separate even if small holes develop, due to natural stratification that limits the leakage flow. Because most

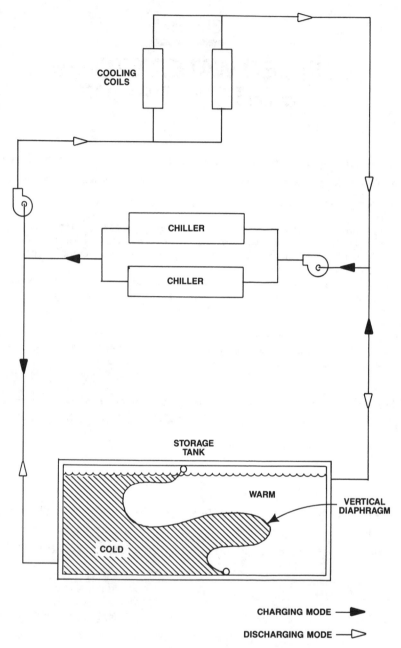

Figure 5-1.   Vertical diaphragm chilled water storage

installations are below grade, however, a means for dealing with support columns is required.   In some cases socks which allow the diaphragm to move freely up and down despite support columns have been used.   Little operational data is available on this technique because it has been applied in buildings only recently.

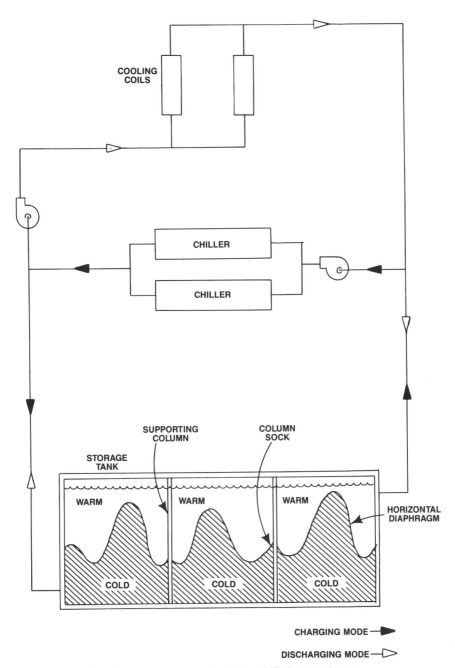

Figure 5-2. Horizontal diaphragm chilled water storage

Multiple Tank

The simplest multiple tank concept comprises a number of tanks connected in series, with piping going from the bottom of one tank to the top of another to avoid "short circuiting." No special effort is needed to inhibit mixing in each tank because, as shown in Figure 5-3, storage efficiency of about 90% can be achieved with 20 tanks in a series, even with complete mixing in each tank (<u>5</u>). Of course, if some degree of stratification were maintained in each tank, the same efficiency could be achieved with fewer tanks.

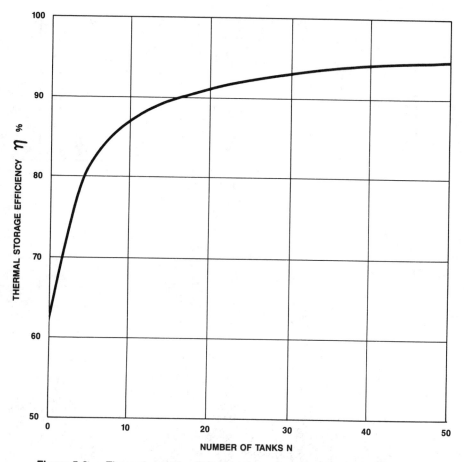

Figure 5-3. Thermal storage efficiency for N tanks in series each completely mixed

Mutiple tanks can also be connected to a piping network with control valves that allow extraction from one tank and return to a different tank (Figure 5-4). In this way the tank that initially is empty can accept the return flow while supply flow is emptying another. This approach completely separates water of different temperatures, yielding 100% storage efficiency. By using a larger number of tanks -- typically six or more -- in this configuration, the unused volume represented by the empty tank is minimized.

The valving required for a multiple tank (tank farm) system makes it a complicated system, and computerized controls usually are needed. These controls, and the

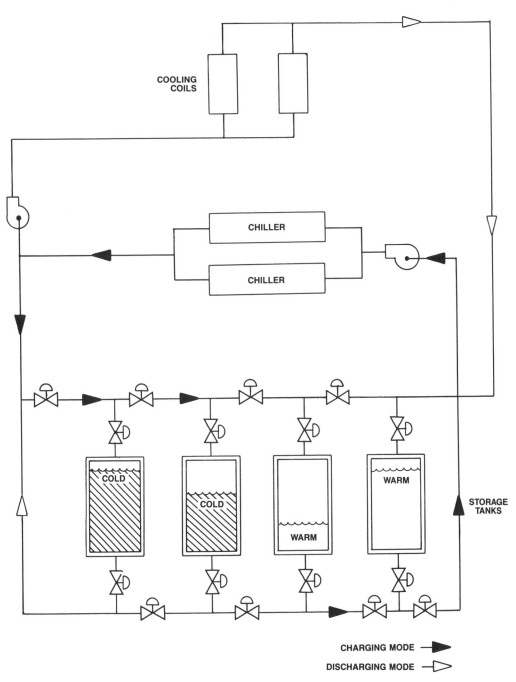

COOLING
COILS

CHILLER

CHILLER

COLD

COLD

WARM

WARM

STORAGE
TANKS

CHARGING MODE ▶

DISCHARGING MODE ▷

Figure 5-4. Empty tank approach to chilled water storage

valving and piping involved, drive up initial costs.   These costs are increased still, more because the walls between tanks must be strong enough to withstand greater static pressure. For example, four small tanks have 2 times more perimeter area than a single large tank of equal volume. When 16 tanks are used, perimeter wall area is increased by a factor of four. Tank farms also tend to be more expensive to maintain given the valving and controls involved.

Despite their drawbacks, many tank farms have been installed, including a 6.3-million-gallon facility built by IBM in Tuscon, Arizona (5). Several other large facilities have been built in Canada and northwestern areas of the United States. Tank farms installed in cooler climates generally are most cost-effective, because modified valve control sequencing permits some of the tanks to be used for hot water storage during winter months. The heat stored can be recovered from the building core, or purchased during off-peak periods.

Stratified Chamber Systems

Stratified chamber systems prevent warm and cool water mixing by relying on buoyant or hydrodynamic effects rather than physical separation. In essence, when cool water is introduced in a tank with warm water, a thin, naturally occuring layer of water -- called a thermocline -- separates cool water from warm water.

Numerous investigations of solar heated storage systems have shown that when warmer fluid is introduced properly a strong relatively thin thermocline may be formed and maintained throughout charging. Recent studies conducted by the University of New Mexico showed much the same results for cool storage systems, indicating that the establishment and maintenance of a thermocline can provide effective separating and thereby enhance storage efficiency (14).

The simplest stratification method is the unbaffled tank system. As shown in Figure 5-5, chilled water is stored in a large unbaffled, unpartitioned tank with inlet and outlet manifolds and distribution laterals designed to ensure low, equally distributed fluid velocities to minimize mixing. Cold water is added or withdrawn through the manifold system at the bottom of the tank; warm water is added or withdrawn at the top.

This is basically the type of stratified system which results of University of New Mexico suggest is the most effective. Although some unbaffled tank systems reportedly have experienced problems, one installed at Stanford University (Palo Alto, California) has been virtually trouble-free (5). A 24-foot-high, 4-million-

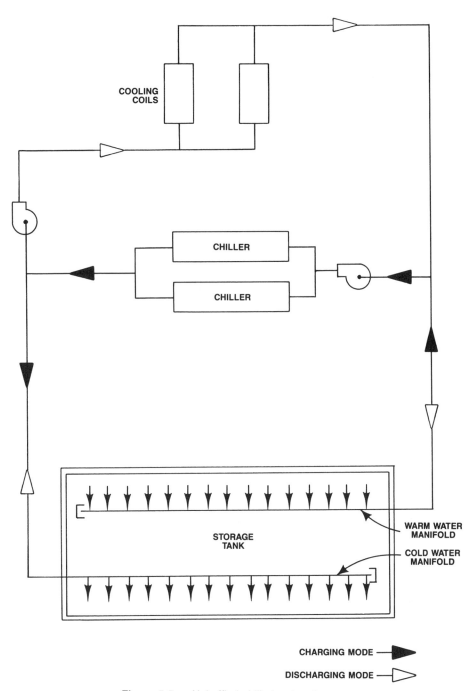

Figure 5-5. Unbaffled chilled water storage

gallon underground tank is employed, and measurements have indicated that mixing is confined to a 3-foot layer (13).

Another form of stratified chamber storage, called a vertically-baffled system, features vertical partitions that divide the tank into several cells or compartments. Two installations in Rochester, New York, use this approach, as do many installations in Japan (5). Numerous design variations are employed, such as the one shown in Figure 5-6. Generally speaking, all operate on the principle of alternating flow paths between cells from top to bottom. Theoretically, this causes the thermal "front" between warm and cool water to move from cell to cell, from one end of the tank to the other, without mixing. Low flow velocities tend to create "dead spaces", that is, regions of a cell where water stagnates and cannot be withdrawn for building cooling. At high flow velocities, mixing occurs due to the turbulence that overcomes the tenuous buoyancy forces separating the warm and cool water. These problems are depicted in Figure 5-7.

These problems of dead space and mixing have been investigated extensively in Japan and Canada. Using model tanks, the Japanese found what they thought to be an optimum method of stratification. Called a ducted vertical baffle chilled water storage system (Figure 5-8), relies on two effects to prevent mixing. First, ducting causes warm water to be introduced at the top of each cell, parallel to the water surface; during reverse flow, cool water always is introduced at the bottom on each cell. Second, low entering velocity and wide entrance ducts minimize mixing of the entering water. (To make the ducts as wide as possible, cells should be much longer than they are wide, and arranged with the ducts on the long side.)

Figure 5-9 shows a typical tank with two cells formed by a single "ducted baffle." (In a multi-cell tank, the ducted baffle arrangement should be the same between each cell.) The design of such a tank begins with determination of supply and return temperatures, to derive ΔT, as described earlier. Applying the Froude number criterion discussed later in this section yields the allowable jet velocity. The jet area then can be found by using this velocity and the desired flow-rate in gallons per minute. Given the flow into the tank, duct length can be determined knowing the entrance velocity. Once duct length is known (assuming the duct runs the length of the cell), an indication of cell size can be found. Cell width should be as small as possible, but large enough so that the combined duct thickness does not reduce the tank's volume more than several percent. For duct walls of concrete blocks, the cell width is on the order of ten to twenty feet.

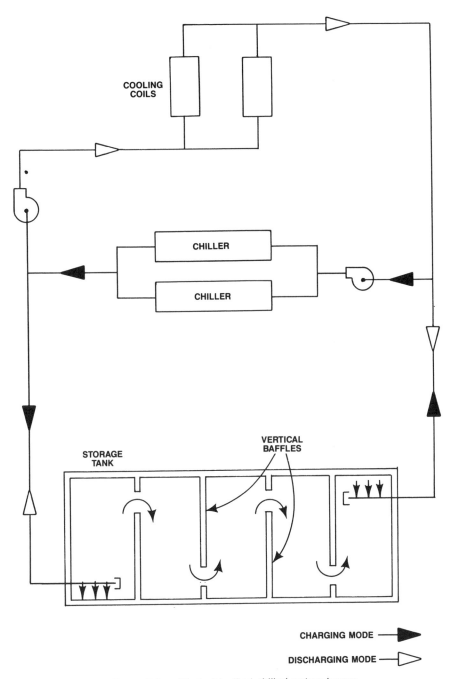

COOLING COILS

CHILLER

CHILLER

VERTICAL BAFFLES

STORAGE TANK

CHARGING MODE ▶

DISCHARGING MODE ▷

Figure 5-6. Vertical baffled chilled water storage

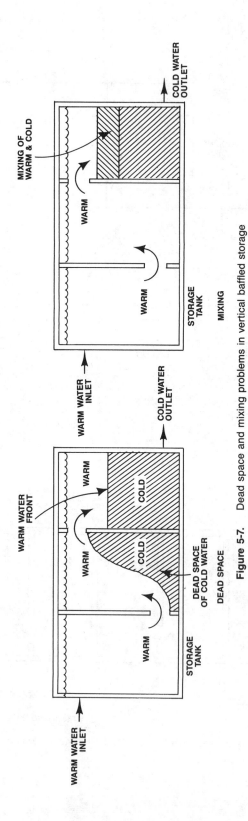

Figure 5-7.   Dead space and mixing problems in vertical baffled storage

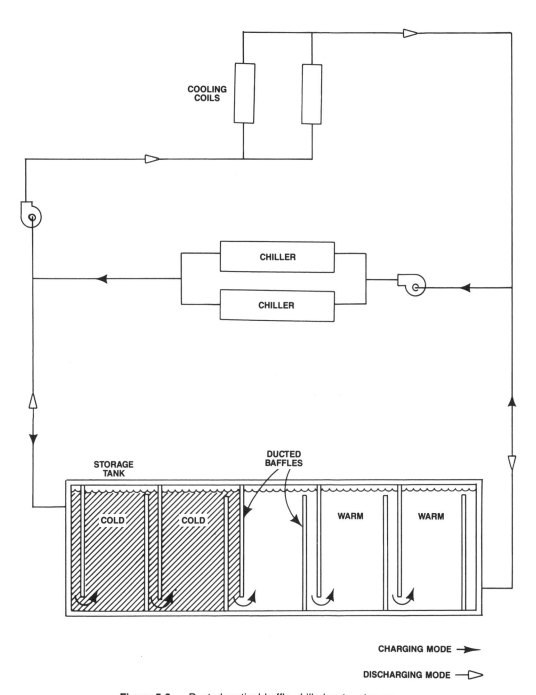

Figure 5-8.   Ducted vertical baffle chilled water storage

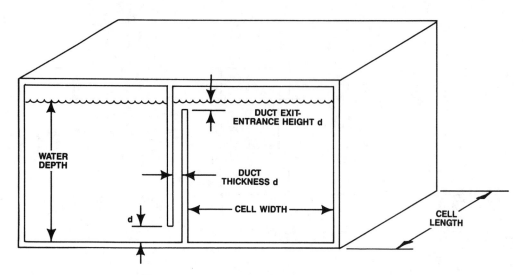

Figure 5-9.   Stratification by ducted baffle

This type of system was analyzed in the University of New Mexico study. Their
results were not as positive as those attained in Japan. It was found that fluid
leaks and heat transfer through internal partitions significantly reduced the
efficiency of the storage tanks (14). Furthermore, the series arrangement of the tank
compartments resulted in additional mixing. It is theorized that the Japanese
obtained superior results because their storage systems involve a very large number
of relatively small tanks in series.

Another stratification scheme -- the suspended vertical baffle system (Figure 5-10)
-- has been used in several Canadian buildings. A patented system, it employs
rubberized sheets hanging from the ceiling of a large tank to form vertical
baffles. These baffles are claimed to cause the water to take a sinuous path from
inlet to output. Vertically oriented manifolds at the inlet and outlet assure low-
velocity, evenly distributed entering water.

The suspended vertical baffle system is particularly appealing from an economic
point of view. The rubberized sheets are comparatively inexpensive, and the
ability to modify the arrangement of the baffling gives flexibility, thus
additional cost-effectiveness. Monitored tests are essential to determine how well
such systems actually work.

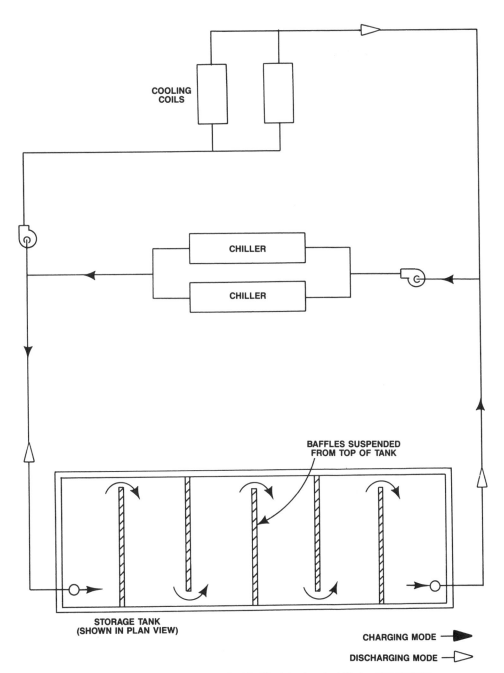

Figure 5-10. Suspended vertical baffle (planform) chilled water storage

STRATIFICATION

A number of factors affect tank stratification and if they are well understood, the designer should have little difficulty in accounting for them properly.

The most important factor affecting stratification is the flow arrangement into and out of the tank. During charging, the chilled water must be introduced at the bottom of the tank and the chiller must draw from the top. During extraction, chilled water must be taken from the bottom of the tank and return water introduced at the top. If this flow arrangement is provided, and precautions discussed below are observed, stratification and storage efficiency can be maximized.

To begin, recognize that the mode of operation changes from charging to discharging, that is, the inflow ports often become the outflow ports and _vice versa_. Therefore, the ports must be shaped for low friction for both directions of flow. Cold inflow/outflow ports are therefore best placed at the bottom, and the warm inflow/outflow ports at the top. To minimize "dead water" regions, ports should be placed as close to the bottom or top as possible.

The velocity and direction of the entering flow stream, referred to as the "jet" affects stratification. For example an upward, high-velocity jet of incoming chilled water is likely to penetrate the lower layers of chilled water and mix with upper, warmer layers. Because high jet velocity will cause turbulence and induce mixing, using low-velocity jets is best. (Maximum allowable jet velocity can be determined as indicated below.) Mixing is also reduced by aiming the low-velocity jets downward for incoming cold water and upward for incoming warm water. As a practical matter, however, success can be achieved with low-velocity, horizontal jets. Mixing around inflow/outflow ports also can be reduced by using calming chambers to damp out the entering jets before the water enters into the main storage section.

The fundamental factor which leads to stratification in a storage tank is the relationship between water, density and temperature. Because colder water is denser, it tends to stay at the bottom of the tank when put there. Thus, if there is a temperature "gradient" in the tank with colder water on the bottom, the corresponding density gradient helps ensure that this situation will continue, unless disturbed by other forces. If there are disturbing forces, such as entry jets, stability of the temperature gradient must be considered. For example, an entering jet of cold water might be propelled by its velocity into a layer in which the local water is warmer than the jet. In this situation, the jet will tend to fall back to a lower level because it is denser than the surrounding water. Limiting the rise height of the jet is termed stabilizing behavior, and is said to be induced by buoyant forces caused by density differences. The strength of the buoyant forces, and therefore the stability of the temperature gradient, is proportional to the magnitude of the temperature gradient. For example, a tank initially at a uniform 60F and in the process of being charged with 50F water will have a stronger tendency to remain stratified than if charged with 55F water.

Note that the strength of buoyant forces in chilled water storage tanks generally is weaker than in hot storage tanks. This occurs because water density approaches maximum at 39.2F and the curve is very flat over the entire range of chilled water temperatures of interest, i.e., 45F - 65F, as shown in Figure 5-11 (5). As such, the density difference between 45F and 65F is only 0.086 pounds/cubic feet, whereas the density difference between 140F and 160F is 0.37 pounds/cubic feet. This makes it more difficult to maintain stratification in chilled water tanks. Much more attention has to be given to minimizing velocities and any other disturbing factors.

Questions about proper test velocity are answered by theoretical and experimental findings which indicate that dynamic forces associated with the jet velocity must be low relative to the buoyant forces associated with the temperature difference between the jet and the still water. A combination of physical parameters which represent the ratio of these two forces is called densiometric Froude number (Fr):

$$Fr = \frac{U}{g\,(\beta\Delta T)\,d}$$

where

U = the entrance velocity of the jet ft/sec
= (volumetric flow rate)/(duct entrance area)

g = gravitational acceleration, 32.2 ft/sec^2

$\beta\Delta T$ = change in relative density for a given temperature difference ΔT between cool and warm water. For chilled water between 40F and 60F has a value of 0.00098 (see Table 5-1)

d = thickness of the jet, ft
= height of duct entrance.

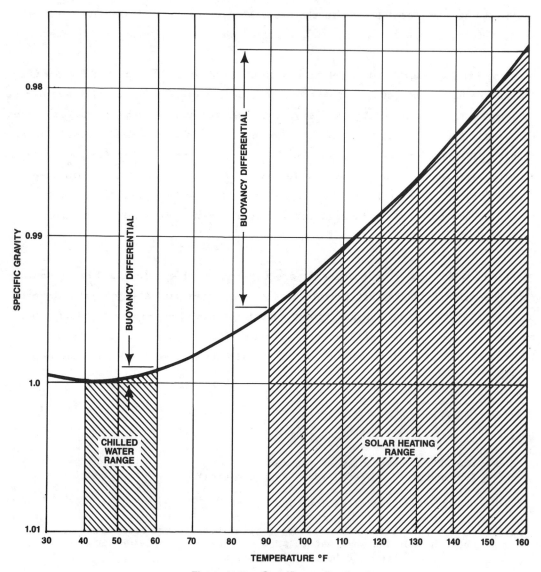

Figure 5-11.   Specific gravity of water

Table 5-1

Relative Density Change of Water

SUPPLY	RETURN		$(\beta\Delta_T = \Delta\rho/\rho)$	
TEMPERATURE (°F)	50	55	60	65
40	0.000275	0.000575	0.000982	0.001475
45	0.000193	0.000490	0.000893	0.001384
50		0.000296	0.000692	0.001185
55			0.000395	0.000878

Figure 5-12 is used to evaluate $\beta\Delta T$ for any two temperatures. For example, if the chilled water temperature is 40F and the return temperature is 60F, the average β for T = (60 + 40)/2 = 50F which is read as 4.9×10^{-5}F^{-1}. The product of $\beta\Delta T = 4.9 \times 10^{-5} \times (60 - 40) = 0.00098$. Table 5-1 gives $\beta\Delta T$ calculated for several combinations of supply and return water temperatures.

Theoretical work done in connection with cooling water discharges into lakes and rivers suggests that the Froude number should be approximately one or less to obtain acceptably low mixing. For example, if the duct height were a half foot, the entrance velocity should be no more than:

$$U = Fr \ \sqrt{g(\beta\Delta T)d}$$

$$= 1\sqrt{32.2 \ (0.00098)(0.5)}$$

$$= 0.126 \ ft/sec$$

For a given water flow-rate into the tank, the area of the entry port determines the jet velocity. That is,

$$U = GPM/(449 \ A)$$

$$A = GPM/(449 \ U)$$

where

U = jet velocity (ft/sec)

A = port area (ft^2)

$449 = GPM/(ft^3/sec)$

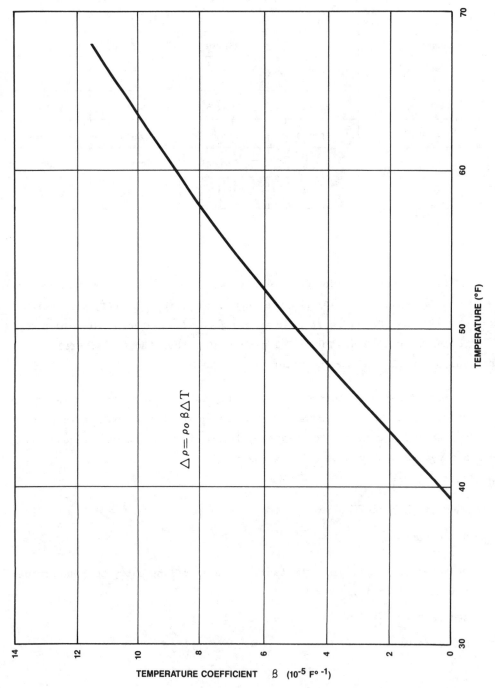

Figure 5-12.   Temperature coefficient of density change for water

To obtain low-velocity jets, port areas must be relatively large. The ideal shape for the chilled water inlet port is therefore a horizontal slot near the bottom of the tank. The ideal shape for a warm water inlet would be a wide slot at the top of the tank. These slots should be along the longest side of the tank and equal in length to the side. The height of the slot should be no higher than necessary to achieve a sufficiently low jet velocity, _i.e.,_ one which gives a Froude number close to 1.0.

STORAGE EFFICIENCY

Because mixing cannot be avoided altogether, storage tanks are less than 100% efficient and allowance for this must be made in sizing.

Stored cooling capacity in a chilled water tank is

$$Q_s = mc_p \ (T_r - T_s)$$

where

m = mass of water in the tank (lbs)

c_p = specific heat of water (1 Btu/lb F)

T_r = return water temperature (F)

T_s = original temperature of water in the tank

Also, if the tank outflow of chilled water is $\dot{m}$ (lbs/hour), the time to replace its contents completely is

$$t_r = m/\dot{m} \text{ hours}$$

If the tank is perfectly stratified, outflow remains at the original tank temperature for the entire period of chilled water replacement. Since this temperature remains, the flow's total cooling effect is

$$\dot{m} \ c_p \ (T_r - T_s) \ t_R$$

By definition of t_R, this is seen to be exactly the same as the stored capacity Q_s. The "storage efficiency" of such a tank is therefore 100%, so that its entire capacity can be extracted in one "filling time."

When mixing occurs, some of the warm return water escapes to the tank outlet sooner than t_R, leaving an equal amount of chilled water still in the tank. As a result, less than 100% of the chilled water can be extracted in the filling time, so storage efficiency is less than 100%. The exact storage efficiency is the percentage of stored capacity which can be extracted in one filling period.

Storage efficiency can be calculated mathematically by assuming a given degree of stratification (or mixing) and dead space. Table 5-2 illustrates extreme situations, and indicates that minimizing mixing and dead space should result in efficiencies of 80% or more (5).

Table 5-2

Chilled Water Storage Efficiency

	DEAD SPACE PERCENTAGE	
DEGREE OF MIXING	0	50%
PERFECTLY STRATIFIED (See Note 1) **(PISTON FLOW)**	100%	50%
PERFECTLY MIXED (See Note 2) **(NO STRATIFICATION)**	63%	43%
COEXISTING PISTON (See Note 3) **AND MIXED FLOW**	82%	49%

Notes:

1. Perfectly stratified or piston flow means that the "mixed layer" is zero, i.e., there is a step change in temperature from the warm to cold layers.

2. Perfectly mixed or no stratification means that the "mixed layer" occupied the entire tank i.e., the entire tank is at a single temperature.

3. Coexisting piston and mixed flow means that there is a "mixed layer" of thickness greater than zero, but less than the entire tank height. In the bottom row of the table, the "mixed layer" is one-half of the tank height.

Note that efficiency (as defined above) does not completely characterize the storage tank performance for purpose of tank sizing. For example, it says nothing about losses due to heat transfer through the tank wall (these usually are negligible) or loss of "usefulness" when mixing raises the outlet temperature above its maximum useful value. Nonetheless, any mixing taking place does reduce its storage efficiency as defined, and it also may reduce the usefulness. Therefore, this efficiency is an indication of loss of usefulness due to mixing.

Storage Tanks

Chilled water storage tanks are constructed from a variety of materials and are available in a number of different configurations. For small installations with capacity requirements less than 40,000 gallons, prefabricated steel and fiberglass tanks are most often used. For larger projects, low-cost cast-in-place concrete tanks are most popular. Tanks that can be integrated into a building's structure, as is the case in a new building, usually are less expensive than tanks sited at a location separate from the building.

Due to the substantial weight of chilled water tanks, they usually are installed on grade, although systems have been installed both below grade and above. In many installations it is worthwhile to employ water storage tanks for storing heated water, too, for recovered heat or heat purchased off peak. Some installations that employ multiple tanks use all of them for cool storage in summer and heat storage in winter. During intermediate seasons, when both heating and cooling might be needed in the building, some tanks store cool water and others store heated water.

In almost all cases, good quality insulation for tanks is readily justified, due to the cost of making and storing chilled water, the large surface areas involved, and long storage times.

SIZING CHILLED WATER STORAGE

Assuming the use of a ducted baffle tank, the first step in sizing the chilled water storage system is calculating the flow rate through the tank baffle. This is a function of the storage discharging rate and the cooling coil temperature differential. It can be calculated using the following formula:

$$m = \frac{\text{discharging rate}}{c_p \Delta T}$$

Using the example discussed in Section 3, having a discharge rate of 5.0×10^6 Btu/hr, and assuming a temperature differential of 18F (*i.e.,* supply at 40F; return at 58F) the flow rate can be calculated as:

$$= \frac{5.0 \times 10^6 \text{ Btu/hr}}{1 \text{ Btu/lb F} \times 18F}$$

$$= 277,777 \text{ lb/hr.}$$

This expressed as a volumetric flow rate is:

$$= 277,777/(62.4 \times 60 \times 60)$$

$$= 1.24 \text{ ft}^3/\text{sec}$$

$$= 556 \text{ gal/min.}$$

The next step is to determine the dimensions of the discharge slot. As discussed earlier, thermal stratification is maximized and mixing is minimized when the baffle discharge Froude Number's near unity. Discharge slots having large aspect ratios (L/H) also enhance stratification. By arbitrarily taking a slot height of 0.5 feet, and the condition Fr = 1.0, the baffle discharge face velocity can be determined:

$$Fr = \frac{U}{\sqrt{g\ (\beta \Delta T)\ d}}$$

Here ΔT is the relative change in density of the chilled water over the design temperature range. Based on 40F supply and 58F returning chilled water, $\beta \Delta T = 0.000732$ as obtained by interpolation from Table 5-1. From this the discharge velocity is found to be:

$$U = 1.0 \ \sqrt{32.2 \times 0.0\text{-}00732 \times 0.5}$$

$$= 0.11 \text{ ft/sec}$$

The discharge slot face area is then given by:

$$A = \dot{V}/U$$

$$= \frac{1.24 \text{ ft}^3/\text{sec}}{0.11 \text{ ft/sec}} = 11.3 \text{ ft}^2$$

Since the slot height has already been chosen as 0.5 feet, the minimum tank length required to give a slot face area of 12.42 square feet is:

$$L = \frac{11.3 \text{ ft}^2}{0.5 \text{ ft}}$$

$$= 23 \text{ ft}$$

The next step is to determine dimensions of the storage tank. Factors influencing this are the required storage capacity, the minimum tank width and the water depth. Ideally, the tank should be square to maximize the tank's volume-to-surface-area

ratio. In this example the tank is located in an open basement mechanical room.
Since there are no architectural constraints other than a ceiling height of 20
feet, the design is directed toward a square tank. Before proceeding, however,
some consideration should be given to how the tank will be partitioned.

Partitions typically are constructed of 6- or 8-inch concrete blocks. In this
example, partitions are made of 6-inch block spaced to give a nominal width of 12
feet. Assuming that a storage efficiency of 95% can be realized using the ducted
baffle arrangement, the required water volume for the storage in the example used
earlier can be determined as:

$$(\text{tank volume})_{\text{ideal}} = \frac{3942 \text{ ton hours x } 12,000 \text{ Btu/ton-hr}}{62.4 \text{ lb/ft}^3 \text{ x } 1 \text{ Btu/lb.F x } 18F}$$

$$= 42115 \text{ ft}^3$$

$$= 315020 \text{ gallons.}$$

As discussed earlier, however, the "mixing layer" volume and dead space detract
from storage efficiency. One way to account for this is by introducing a "storage
efficiency", η_s, so that:

$$(\text{tank volume})_{\text{actual}} = \frac{(\text{tank volume})_{\text{ideal}}}{\eta_s}$$

Assuming a well-designed system, η_s, theoretically can be as high as 90 to 95%.
Thus for the above example, the required storage capacity is:

$$(\text{tank volume})_{\text{actual}} = \frac{42115}{0.95} = 44,332 \text{ ft}^3$$

Using an average water depth of 14 feet and a square plan, the required interior
side length is:

$$s^2 = \frac{44,332 \text{ ft}^3}{14 \text{ ft}}$$

$$= 56 \text{ ft.}$$

The inside dimensions of the tank thus are 56 ft x 56 ft x 14 ft. However,
dimensions should be increased slightly to allow for the wall partitions in the
tank. If the tank is divided into five cells (normally cells should be 10 to 20
feet in width for good stratification), and each partition is six inches, the

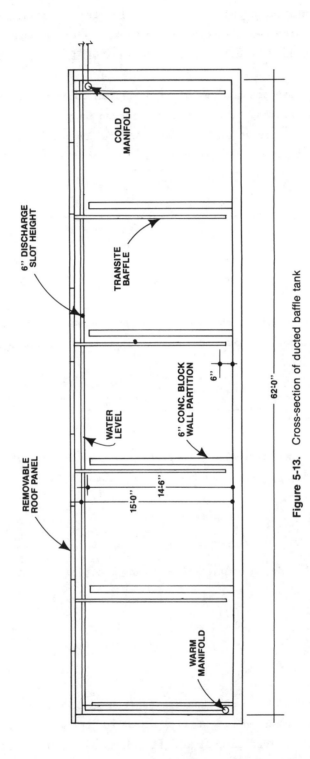

Figure 5-13. Cross-section of ducted baffle tank

storage tank's inside dimensions will be 56 x 58 feet. Thus, a storage tank having these dimensions will give the required water volume and partition spacing and still be nearly square, as shown in Figure 5-13. The perimeter walls of such a tank are constructed with 12-inch reinforced concrete with a waterproof polyethylene membrane and a 3-inch gunite interior face. The walls are insulated and extend six inches above the water surface. This provides sufficient freeboard to hold the chilled water in the piping loop and to accommodate water level fluctuations that occur as water is pumped through the tank. The tank roof is composed of removable, insulated panels which are suspended in a steel frame on top of the tank. The roof's purpose is to reduce heat transfer to the chilled water, keep debris out of the tank, and provide access to tank compartments.

As indicated in Figure 5-14, the tank's partitions must have a flow passage gap or slot at the top, and there must be a thin, nonstructural flow guide sheet extending from the top of the tank, offset from the partition. The flow guide sheet ensures that flow from one compartment to the next proceeds without mixing. The offset distance should be equal to the slot height (six inches in this case). The flow guide sheets can be of any material that will not corrode; transite is one choice.

Tank manifolding is accomplished using perforated tubular headers behind transite baffles at opposite ends of the tank. As shown in Figure 5-14, cold water is drawn from the cold manifold and warmed water is returned via the warm manifold. During charging, flow is reversed; cold water enters from the cold manifold and warm water is removed at the warm manifold. The baffles provide a calming region for the manifold flow, to minimize mixing.

EQUIPMENT ARRANGEMENT & PIPING

A chilled water storage equipment arrangement concept that has been used successfully is shown in Figure 5-15A (12). As indicated, storage can feed chilled water to building pumps by day, can be regenerated at night by the chiller, when the building circuits may be idle, and it can supplement an undersized chiller by helping it through the day. Note how even a minor variation in chiller placement defeats the flexibility shown in Figure 5-15A. For example, the storage shown in Figure 5-15B can be recharged only by reconditioning the entire tank volume. Even if a small percentage is drawn for one day, the preparation of storage for maximum load on the next day requires that all the water be passed through. Another problem with the circuit in Figure 5-15B is its inability to effect an electric demand limitation strategy. The chiller is poorly located to share cooling load with storage and is more likely to be turned on at full capacity when storage runs out.

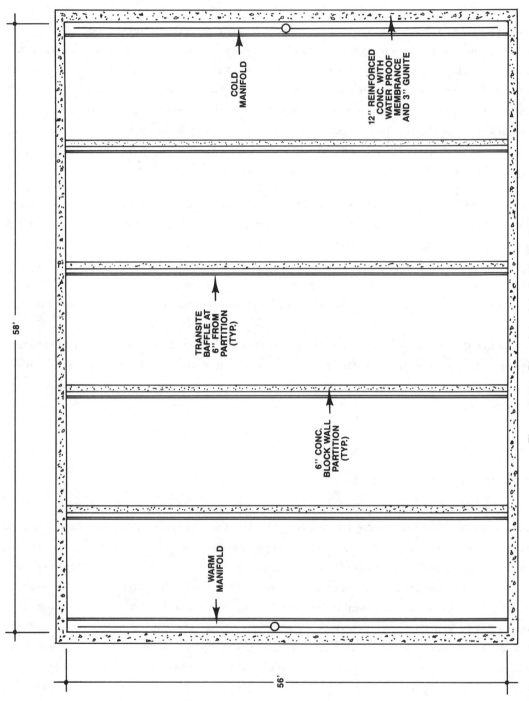

Figure 5-14.   Plan view of ducted baffle tank

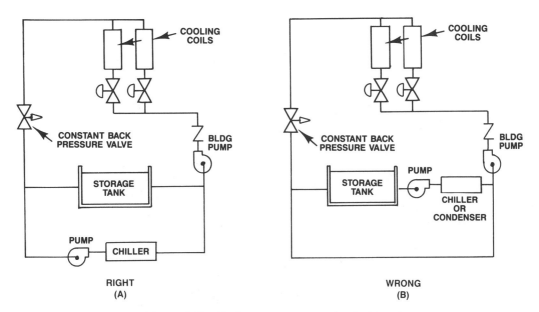

Figure 5-15.   Equipment arrangement and piping

Using the basic concepts embodied in Figure 5-15A, a schematic of the arrangement for the 15-story office building example discussed in Appendix A and earlier in this section is shown in Figure 5-16. Note that a heat exchanger has been used in this illustration to offset hydrostatic pressures, as discussed below.

Direct Pumping vs. Heat Exchanger

Figure 5-15A illustrates direct pumping of building circuits, an approach used typically in Japan. Figure 5-17 illustrates the heat exchanger concept used typically in North America (13).

The perceived problems with direct pumping are the waste of energy to overcome static head, the uncertainty of hydraulic pressures and water conditioning in an open, variable flow system, and the novelty of hardware such as constant back pressure valves, used to prevent flooding of storage tank and air entry to building circuits. In fact, however, the hardware for open circuit pumping has been well tested and the choice of whether to use a heat exchanger should proceed with the usual engineering analysis of pros and cons.

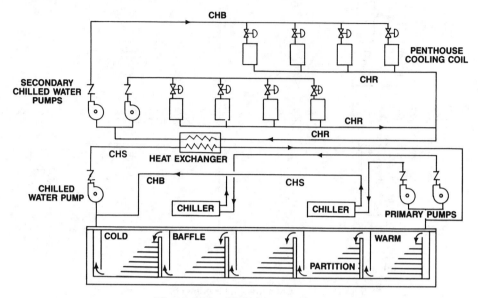

Figure 5-16.   Chilled water storage system

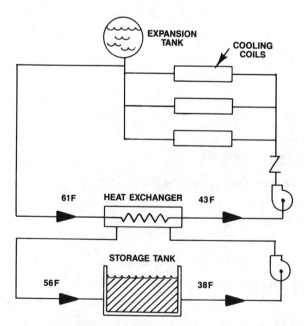

Figure 5-17.   Bottom side storage with heat exchanger

The heat exchanger pros consist of the isolation of building circuits from the presumably contaminated water in open storage and the elimination of static head pumping energy. The cons involve increased energy for pumping through the resistance of both heat exchanger circuits and the increased chiller energy to generate a lower chilled water temperature to compensate for the heat exchanger approach ΔT. Another con is the cost of the heat exchanger and its extra pump, including the space they occupy and the extra maintenance they create.

The energy trade-off is illustrated on Figure 5-18. For a single pumped circuit, the building should be 15 stories high before the static head pumping energy exceeds the extra chiller energy for overcoming the typical ΔT of the heat exchanger. Before the heat exchanger amortization is equalled by direct pumping energy, however, the building may have to be another 15 stories higher, as shown on Figure 5-18 (<u>13</u>).

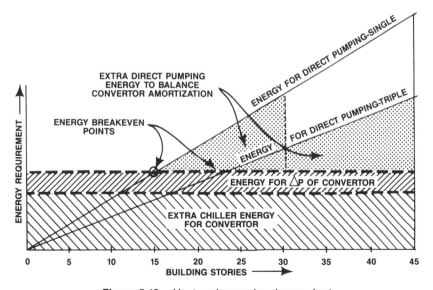

Figure 5-18. Heat exchanger breakeven chart

The Japanese justify direct pumping on buildings up to 45 stories by splitting the pumping into three or more separate circuits, as shown on Figure 5-19.

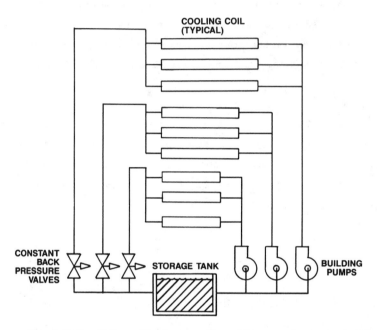

Figure 5-19   Bottom side storage with multiple pumping to the building

Two-way vs. Three-way Valves

Many designers have abandoned the three-way valve as a control mechanism for cooling coils. They rely instead on less expensive two-way throttling-type control with vari-speed equipment to control pump output and eliminate pressure fluctuations, and reduce energy consumption. Note, precautions must be taken so that chilled water flow through chiller does not reduce below the chiller package manufacturers recommendations. Below this range (which is approximately 35% of nominal rating) efficiency falls off dramatically and freezing of tubes is likely to occur. Further, if the full load flow is not kept within the design parameters, the higher fluid temperature differences cannot be accomplished, and increases in system pressure differential will result in excessive noise in automatic control valves.

Figure 5-20 depicts the leaving water temperature from a chilled water coil at various loads, using a two-way and a three-way control valve. Note that the range is maintained with a two-way valve down to 50% load (range is less critical below

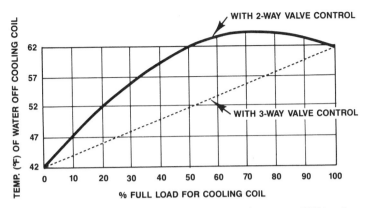

Figure 5-20. Coil performance with 42F entering water; 55F leaving air

50%), while the three-way counterpart erodes the range immediately as the load subsides (13).

As a result, three-way valves should never be used on storage installations. They are more expensive and they contradict the basic storage principle of maintaining temperature range.

Cooling Coil Surface

The chilled water temperature range most commonly supplied for conventional chilled water systems is 8 to 12F. For chilled water storage installations, it is now conventional to select a range between 18 to 24F. If the range is not achieved, the portion of cooling load furnished by storage will fall short.

It is entirely likely that most conventional cooling systems never achieve the chilled water range for which they were designed, yet still provide satisfactory performance because of safety factors included in calculation of load and sizing of the air handling plant. When storage is used, chilled water range must be achieved or the plant will deplete its cooling capacity before the day is over.

For cooling coils, inability to achieve water range for a given cooling load indicates an improper coil selection.

Maintenance of Temperature Range. Although a chiller may be expected to produce water at a constant design temperature, this becomes impossible if chiller capacity is being restrained by a demand-limiting strategy or if the chiller must accept water at room temperature on original pull-down. Design water temperature can be achieved by using a bypass on the building circuit for return temperature and a bypass on the chiller for supply temperature (Figure 5-21).

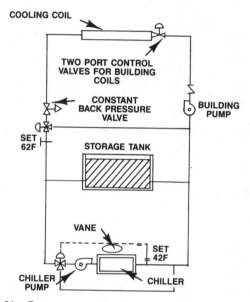

Figure 5-21. Bypasses to maintain temperature range in storage

Return water temperature can also be kept relatively constant through use of two-way throttling valves on the cooling coils, but building loads can fall below 50%, where the range cannot be maintained. Some circuits, such as one with a group of fan coil units, may be incapable of producing the chilled water range.

Connections to a Multi-Compartment Tank

Most designers prefer to divide a large storage into several compartments. Among other benefits, this permits one module to be drained for repair without losing the entire storage. In cases where a fraction of the storage is convertible to heating, there must be at least one partition in the tank.

When water is withdrawn from one compartment at a time, controls are complex and expensive. When water is drawn from two or more modules operating in parallel, another problem may develop. As shown in Figure 5-22, if the difference in resistance among the water paths in or out of a compartment exceeds as little as one foot, there will be a foot of difference in water level in the compartments. A drop in one enforces a rise in others. If the water rises to the overflow, water

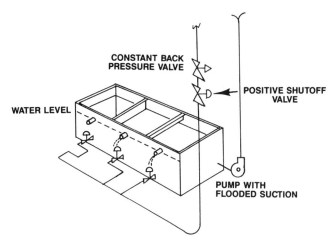

CONSTANT BACK
PRESSURE VALVE

POSITIVE SHUTOFF
VALVE

WATER LEVEL

PUMP WITH
FLOODED SUCTION

CAUTIONS IN STORAGE TANK DESIGN:

1. PUMPS TO HAVE FLOODED SUCTION.
2. CONSTANT BACK PRESSURE VALVE TO HAVE AUXILIARY POSITIVE SHUTOFF VALVE.
3. COMPARTMENT HEADERS TO BE OVERSIZED WITH ALL CONNECTIONS IDENTICAL. AVOID UNEQUAL PIPING AS ILLUSTRATED

Figure 5-22. Water drawn from two or more modules operating in parallel

will be wasted, creating an expense both for the water and corrosion treatment. The antidote for this potential problem is to oversize the pipe headers feeding the compartment so they act as a plenum. The connections should then be arranged to be as identical as possible.

Two other items of importance are illustrated in Figure 5-22. One is to ensure that the pumps have flooded suction, to prevent complicated pump service and operation. The Japanese employ turbine type units in order to retain the pump motors above the tank top for service. Others arrange a pump room alongside the tank and use conventional pumps. The other item concerns the constant back pressure valve. Most models are unable to hold tight at zero flow. Problems caused by this are relieved by automating an isolating valve to act as a stop valve when the circuit is inactive.

Filter

The large volume of water in a cool storage system, and the large surface open to atmosphere above the typical tank, make maintenance of satisfactory water conditions a critical concern. Swimming pool filters are the least expensive and most effective type of equipment for maintaining good water quality. The recirculation rate should be at least one turn-over in two days (13).

Figure 5-23 illustrates three possible filter locations. The separate filter pump
is best, not only to permit constant circulation, but also to enable the pump to be
selected for the resistance of the filter rather than some storage circuit. The
bypass location on the building pump is entirely out of the question since the
filters seldom require building heat and the building pumps are usually time-
cycled. The bypass location on the primary pump is better, but still subject to
time cycling.

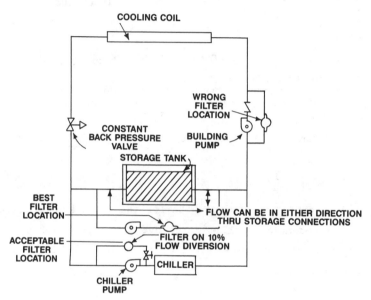

Figure 5-23. Use of filters

Tank Liners

Water seepage into or out of storage generally is not an urgent matter, but leakage
is unsightly and is a potential source of water channelling that could undermine
the structure. Many poured-in-place storage tanks are operating satisfactorily
without liners, while several have been salvaged only by adding liners later.
Since liners add approximately $2.50/sq.ft. to the cost of tanks (13), it is felt
that they should be added later if the system is unsatisfactory without them. The
owners should be advised of this potential, and should be encouraged to establish
contingency funding for it.

RECLAIMING WASTE HEAT

A typical refrigeration machine with a water-cooled condenser rejects approximately 15,000 Btu/h for each 12,000 Btu/h of refrigeration. A significant amount of this heat can be captured and used for heating service hot water, space heating, preheating boiler make-up water and tempering of ventilation air. A commonly employed method for heat recovery involves the use of a double bundle condenser system.

A double bundle condenser is constructed with a building water circuit and a cooling tower circuit enclosed in the same shell. Hot refrigerant gas from the compressor is discharged into the condenser shell where its heat is absorbed by either one of the water circuits or by both simultaneously depending on the requirements of the system at a given time.

The condenser is split into two independent hydronic circuits to prevent contamination of the building water system with cooling tower water, which may contain dirt and chemicals.

Since in a chilled water storage system most of the refrigeration equipment operation occurs during off peak hours when the building is unoccupied, sufficient heat may not be available directly from the refrigeration plant during on-peak hours or other occupied periods. If the quantity of recovered heat is not adequate and heat recovery is cost-effective, it will be necessary to recover heat during off peak hours and allow for additional storage capacity for heating.

CONTROL OF CHILLED WATER STORAGE SYSTEMS

Chillers, pumps, and valves are controlled in that manner required to implement the basic operating mode involved. Fundamental control decisions determine when the chiller will operate and when storage will be charged and discharged.

One control method is to run the chillers according to a prescribed schedule, i.e. capacity vs. time of day, with override from temperature sensors in the storage unit; fully charged storage should inhibit further chiller operation. Chiller output always goes into the storage unit, and supply chilled water for the building cooling coils is always taken from the storage unit to meet the instantaneous load. Whenever chiller output exceeds building cooling demand, a net increase in stored chilled water results. When the chiller is scheduled off, or when its output is lower than demand, there is a net reduction in chilled water stored. However, when high-rise buildings and unpressurized tanks are involved, variable- or multiple-speed pumps are advised. This is so because, with an unpressurized system, the elevation head is lost to the pressure reducing valve, creating a significant pumping energy cost. By reducing flow, perhaps under control of return water temperature, pumping costs can be reduced at low loads.

Chapter 6

REFRIGERATION SYSTEM COMPONENT SELECTION

Careful selection of all mechanical refrigeration system components is essential to proper and efficient operation of both the ice storage and chilled water systems. The following discussion provides information on the various types of compressors, evaporators, condensers, and refrigerant feed devices that are available and which of these choices are preferred for application with ice storage or chilled water systems.

COMPRESSORS

The refrigerant compressor is a mass mover which transfers refrigerant vapor from a low-pressure area at evaporating conditions to a high-pressure area at condensing conditions. On the low-pressure side, the compressor may see evaporating temperatures as low as 15F average when making ice, and as high as 35F to 40F when chilling water or 45F to 50F when cooling air.

On the high-pressure side, condensing temperatures may range from a high of 125F in heat pump operation to a low of 65F when condenser heat is rejected during intermediate seasons. Due to the range of operating conditions, minimizing thermal lift to reduce compressor power requirements is a prime consideration. Thermal lift is analogous to the pressure differential of a fan or pump. Compressor thermal lift is defined as the temperature differential between discharge (condensing) and suction (evaporating) conditions. As thermal lift is increased, so is the required compressor brake horsepower. Therefore, in the interest of minimizing compressor power requirements, thermal lift should be minimized by designing for lowest possible condensing temperatures and highest possible evaporating temperatures. Figure 6-1 shows a range of capacities of different types of compressors used with efficient cool storage systems.

A centrifugal or turbo compressor is most commonly used on medium and large water chilling systems for air conditioning, although some with slight modifications may be used for ice making as well. It is basically a constant head machine. Thus, if it is operated beyond its design range, surging or hunting occurs, accompanied by noise, vibration, and heat. Most machines on the market use R-11

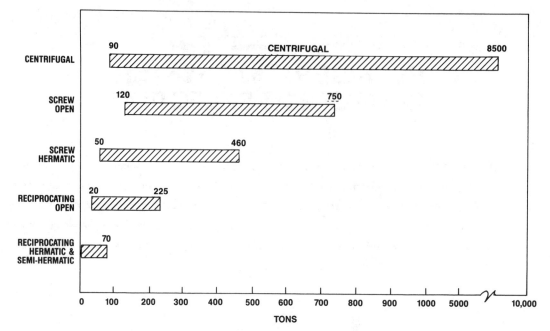

Figure 6-1.   Approximate range of capacities for compressors

which is a low pressure refrigerant requiring a purge unit to prevent air entrainment.

Centrifugal compressors are the preferred choice in chilled water storage systems with capacities above 100 tons. Open-drive centrifugal compressors and chillers are available in capacities ranging from 90 to 1,250 tons. Large field-assembled, open-drive centrifugal compressors and chillers are available in capacities ranging from 700 to 5,000 tons. Multiple stage compressors extend the range up to 10,000 tons.

Hermetic or Semi-hermetic compressors are not used in energy efficient ice storage systems. For example, a typical four-cylinder semi-hermetic compressor of about 40 CFM displacement (15 hp) operating at 100F condensing and 20F evaporator, with 65F gas temperature at the compressor suction valve will use 10% more kW/ton than an open compressor of the same size. When 80 CFM displacement (30 hp) compressors are used, the semi-hermetic compressor uses 20% more kW/ton.

Likewise, an air cooled condensing unit at 115F condensing with a semi-hermetic compressor may use 150% more kW/ton than an open compressor with an evaporative condenser operating at 95F condensing temperature. This type of compressor is not available with capacity unloading due to suction controlled motors. It is not adaptable to an economizer cycle as are centrifugal and rotary screw compressors and therefore is not as efficient.

The **open reciprocating compressor** is generally applicable to systems requiring a compressor of 20 to 300 HP. This type of compressor usually has capacity control by loading or unloading cylinders. As such, the reciprocating compressor has a step response to load variation. Nonetheless, this type of compressor accommodates well the varying load characteristics of water chilling without storage. When applied to ice or water storage, the unloading capacity control is not needed. Reciprocating compressors are best suited for small to medium ice builder systems where they can be matched on a one-for-one basis with an ice builder. The largest factory-assembled ice builders require a maximum compressor size of about 150HP.

For larger installations where capacity requirements would necessitate a multiplicity of reciprocating units, the rotary screw compressor would be the logical choice.

The **rotary screw compressor** is highly efficient and reliable. Because its rotors are separated by an oil film, it exhibits less wear than reciprocating machines. In addition, the screw compressor features efficient, surge-free, infinitely variable (slide valve) capacity control from 100% to about 10% of full load. Rotary screw compressors are available in sizes ranging from 100HP to 1500HP.

Packaged Chillers

Packaged chillers employing reciprocating, centrifugal or screw-type compressors are available in a variety of sizes. Their cooling capacity in the manufacturers literature is most commonly based on standard design conditions specified by the Air-Conditioning and Refrigeration Institute.

Packaged water chillers are normally rated with a difference in chilled water temperature between entering and leaving the cooler of 20F. When the temperature difference is increased -- as is the case in chilled water storage design -- the water flow quantity decreases, i.e., with an increase from 10 to 20 degrees, the flow rate is 50% of maximum and minimum water flow ratings. Manufacturers generally do not recommend flows beyond those values. As the flow rate becomes less, the tube velocity decreases. At lower tube velocities, the heat transfer rate decreases exponentially resulting in increasing capacity loss with greater likelihood of water freezing in a tube. The freezing potential results from a combination of low tube velocity and lowering of refrigerant pressure (and temperature). The problem is further compounded when, due to 2-way valve throttling, the flow reduces in response to a reduction in system demand.

A review of packaged water chillers of two manufacturers in the range of 30 to 100 tons shows that the recommended tube velocities is between 3.3 to 10 and 2.4 to 12 (feet per second), respectively, allowing a maximum reduction of waterflow from nominal (10F) of approximately 51 to 61% and 44 to 29%, respectively.

Compressor Selection and Sizing

The compressor selection is based on knowledge of required capacity, average suction temperature, and the condensing temperature. The compressor size is determined by dividing the ton hours of thermal storage by the number of hours the system is operated each day. It is essential to select a compressor that at full load will just recover the rated storage in the available time. If an oversized compressor is used, it will be less efficient, will operate at lower suction temperature, and will require more kW per ton. In all cases, manufacturers' tables and operating specifications should be consulted.

Matching Multiple Ice Builders With Multiple Compressors

Multiple ice builders commonly connected with multiple compressors present the flexibility of better ice management, and give better efficiencies when some of the compressors are not needed during lighter cooling load seasons. The larger evaporator surface presented to fewer compressors allows both higher evaporator temperatures as well as lower head temperatures because of the extra condensing capacity available. As such, less thermal lift means less energy per ton hour. As a rule of thumb, each 1F lower head yields 1 percent more capacity.

A compressor start/stop control is an important consideration when multiple compressors are involved. Ice thickness controls are continuously used for this purpose. Others include suction pressure control, or microprocessor control with telephone links to remote load management auditing systems. The need for a common oil return system also is a concern. In the case of reciprocating compressors with low-pressure crankcases, an oil still/receiver with low side oil float valves on the compressors is required. (Figure 6-2). For multiple screw compressors, a simple bleed from the pumped liquid delivered to the suction line has been effective (Figure 6-3).

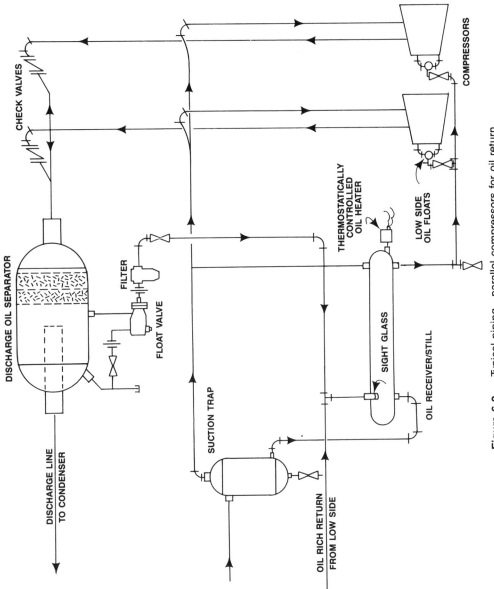

Figure 6-2.   Typical piping - parallel compressors for oil return

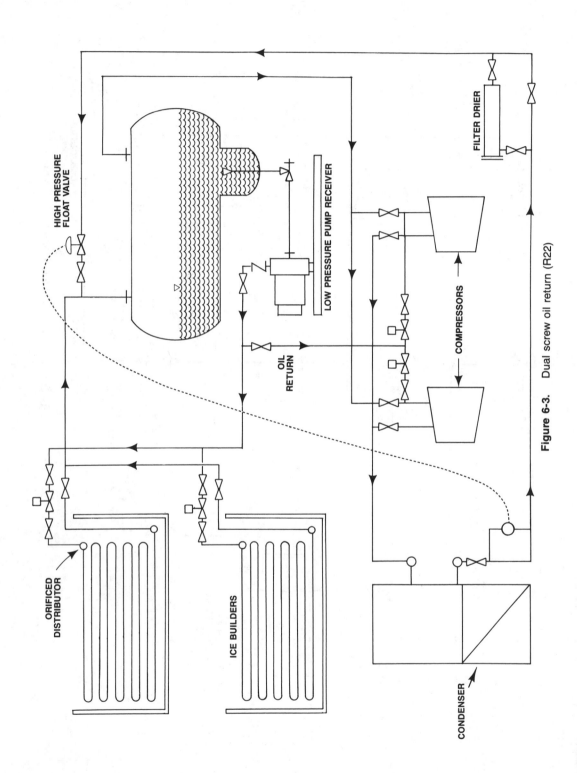

Figure 6-3. Dual screw oil return (R22)

CONDENSING EQUIPMENT

The selection of the heat rejection equipment (condensing equipment) has a significant impact on compressor and system power requirements. Three basic types of condensers are used: air-cooled, water-cooled, or evaporative. Each type has advantages and disadvantages which should be evaluated against the specific design criteria in order to determine the best choice for each project.

Air-Cooled Condensers

An air-cooled condenser removes heat from a refrigerant and then condenses the refrigerant by blowing air over an extended surface coil through which the refrigerant vapor is circulated. The latent heat of the refrigerant is removed by sensibly heating the air, so the condenser capacity is determined by the ambient dry bulb temperature.

The primary advantage of air-cooled condensers is their simplicity; they consist of a coil, fan system, and enclosure. A secondary advantage is the low maintenance, because of the elimination of water, scaling, treatment, corrosion etc. However, they contribute to less efficient operation when compared with water-cooled or evaporatively cooled condensers, because of their higher condensing temperatures (water or evaporative temperatures usually are 15 to 30 degrees lower than dry bulb temperatures). Also, they require large fan horsepower to move large volumes of air, and they take up more space than water-cooled condensers. As such, air-cooled condensers should be considered for ice storage only in smaller sizes (up to 20 ton) and only in moderate climates.

Water-Cooled Condensers

The water-cooled system includes a shell-and-tube condenser, cooling towers, or once-through water source and condenser water pumps. Heat rejection takes place in two steps: The refrigerant is condensed by the water flow in the condenser, and then the heat is rejected to the atmosphere as the water is cooled in the cooling tower. Since this arrangement utilizes the evaporative cooling principle (in the cooling tower), it rejects heat in a relatively efficient manner. The total energy required to (pump horsepower plus cooling tower fan horsepower) reject the heat is less than for a air-cooled condenser at the same condensing temperature. The equipment requires less space, too, and generally will cost less.

Typical design conditions are 75F to 78F ambient wet bulb, 85F entering condenser water temperature, 95F to 100F leaving condenser water temperature, and 100F to 105F saturated condensing temperature.

Due to lower level heat sink (ambient wet bulb temperature) and closer approach via better heat transfer, compressor condensing temperature is substantially lower, improving compressor kW/ton. Water consumption and treatment are occasionally cited as a disadvantage of water-cooled systems, but energy considerations usually have greater impact. Also, the water cooled condenser/cooling tower approach can be the best solution on installations where the heat rejection equipment is remote from the compression equipment, and local codes limit or prohibit the installation of refrigerant lines in or near occupied spaces.

Evaporative Condensers

The evaporative condenser combines a water-cooled condenser and cooling tower in one piece of equipment. It eliminates the sensible heat transfer step of the condenser water, permitting a condensing temperature within 15F of the design wet bulb, resulting in compressor horsepower savings of 10% or more over water-cooled condenser/cooling tower systems, and more than 30% over air-cooled systems. Fan horsepower is comparable to water-cooled condenser/cooling tower systems, and is about one-third of an equivalent air-cooled condenser. Design conditions with evaporative condensers are usually 90F to 95F saturated condensing temperature with an ambient wet bulb temperature of 72 to 78F.

Because the flow rate of water in an evaporative condenser need only be enough to thoroughly wet the condensing coil, water flow rate and head are reduced, pumping horsepower is only about 25% of that required by a water-cooled condenser/cooling tower system.

The evaporative condenser combines the cooling tower, condenser surface, water circulating pump, and water piping in one assembled unit, thereby eliminating the cost of shipping, handling and installing separate components. And since the evaporative condenser utilizes the efficiency of evaporative cooling, less heat transfer surface, fewer fans, and fewer fan motors are required, resulting in material cost savings of 30 to 50% over a comparable air-cooled condenser.

By combining the condensing coil and cooling tower in one unit, the evaporative condenser saves valuable space. It also requires about 50% of the surface area of a comparably sized air-cooled condenser, since it requires only about one-fourth of the airflow.

Considering the lower first cost, space savings, and especially the minimum energy requirements, evaporative condensers should be the best choice in condensing equipment for most ice storage systems.

Heat Recovery Equipment

Since the refrigeration system must reject heat to some condensers it should
always be a consideration to utilize that heat whenever possible. Choices for
useful instantaneous heat recovery condensers include refrigerant-to-air heating
coils and refrigerant-to-water heat exchangers for building heating water and/or
domestic water. For heat recovery, a maximum condensing temperature of 125F is
recommended.

REFRIGERANTS AND REFRIGERANT FEED METHODS

Three general types of liquid refrigerant feed systems are in use with ice storage
systems: direct expansion, gravity circulation and liquid overfeed systems.

Direct Expansion

Direct expansion refrigerant feed is best suited for R-22 applications where a
single ice builder is matched with a single dedicated condensing unit with a fixed
charge of refrigerant, a shell and coil (suction trap) and one or more orifices,
thermostatic, or automatic expansion valves feeding distributors to provide flow of
liquid to each pipe coil in the ice storage.

The usual trim for R-22 systems (in addition to the expansion device and the
suction trap) may include a liquid solenoid valve controlled by ice thickness
control, an oil return bleed, a liquid filter-dryer, a suction strainer and a
crankcase heater for compressor protection.

Systems whose liquid charges are 20 times more than the crankcase oil charge should
also be fitted with discharge oil separators to prevent nuisance shut-downs caused
by oil migration. A typical direct expansion feed arrangement for a simple system
is shown in Figure 6-4.

The use of a liquid coil in the suction trap makes it possible (with R-22) to slop-
over up to 20 percent of the liquid flow. This overfeed is used to precool the
feed to the orifice or expansion valve. This assures a "wetted" coil through its
full length, and reduces velocity in the coil by about 20 percent. This reduced
velocity reduces the pressure drop by one-third, which -- in turn -- allows a
higher evaporating temperature. A higher evaporating temperature improves the
system capacity, so that these simple systems harvest almost as much ice as the
more complex flooded and overfed systems.

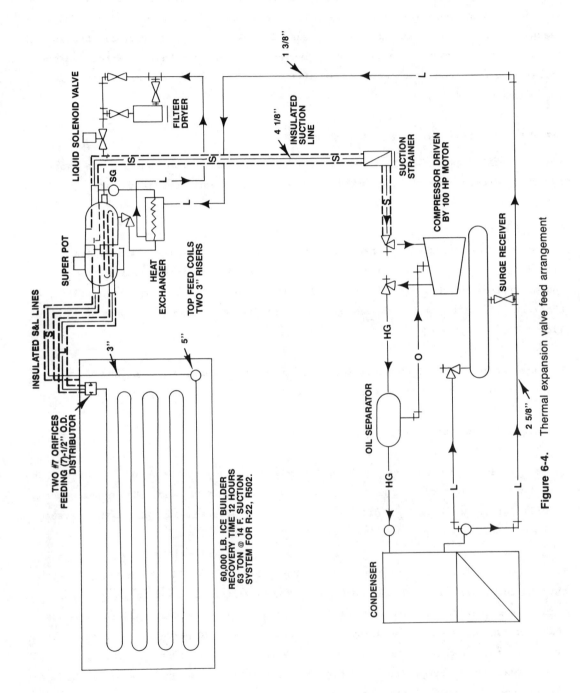

Figure 6-4.   Thermal expansion valve feed arrangement

By top-feeding the pipe coil, the operating charge is reduced to about one-third the volume of the pipe coil. By contrast, bottom-fed coils require from 2 to 3 times the charge. Since halocarbon refrigerants are expensive, minimizing the charge reduces potential dollar losses caused by a leak.

Thermostatic expansion (TX) valves also have been used, usually one for each pipe coil or pair of coils. This construction requires a separate suction pipe for each thermal bulb. Since some superheat is required to actuate the TX valve, these systems run 5 to 7 degrees lower suction temperature than overfed systems and produce about 10 to 15 percent less ice with a given condensing capacity.

An electric TX valve operates on a thermistor at the liquid-vapor interface, or 0 degrees superheat. Demonstration projects with such valves (using shell and coil suction traps) show promise of performance almost as efficient as more complex flooded or overfed systems.

While individual TX valves are preferred for best distribution of liquid refrigerant, multi-outlet distributors may be used effectively in combination with shell and coil suction traps.

Gravity Circulation

Ever since the ice builder was developed in the 1930's the standard ammonia control has been a gravity-circulating thermal-syphon (Figure 6-5). This device uses a surge drum mounted above the ice tank, with a volume between 25 and 100 percent of the pipe coil, depending on the design of the circuits and the needs of the plant to which it is connected. If the compressors are not protected with an auxiliary accumulator, the surge drum volume should be 100 percent of the coil volume. The drum is fitted with a liquid feed float valve to maintain a liquid head above the ice bank, feeding the bottom of the coil through a liquid leg. The thermal action of the boiling ammonia in the coil creates vapor bubbles which reduce the density of the liquid in the pipe coil and in the gas leg. As a result, the liquid leg head forces the boiling liquid in flow multiples as high as 50 times the evaporating rate. The resulting heat transfer rates are the best that can be achieved as long as the forcing head is more than the resulting "heat-head" of the coil. When too much compressor is used or the circuits are too long for heavy loading, such as occurs when all of the ice is gone, the thermal syphon stops, the liquid backs into the drum and the coil becomes "gas-bound." Thermal syphons can be upset by pressure variations caused by changes in liquid feed make-up to the surge drum.

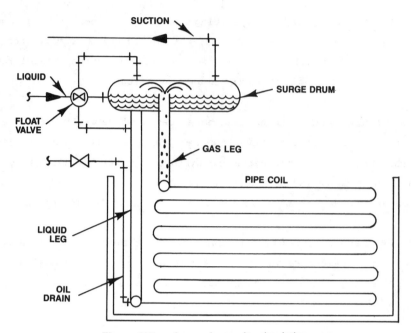

Figure 6-5.   Ammonia gravity circulation

<u>Thermal syphons should not be used with halocarbons,</u> unless aided by liquid injection directly into the coil, because of the relatively large liquid make-up rates with halocarbon refrigerants.

<u>Injection-Aided Ammonia Circulation.</u> By installing the liquid feed into the liquid leg below a check valve, it is possible to force higher tonnage capacity into over-burdened or undersized ice builders. This is particularly useful with ammonia bottom-fed coils in plants where ice water needs have grown beyond the original design (Figure 6-6).

<u>Injection-Aided Refrigerant 22 Circulation.</u> For R22, an injection system for top feed coils has been widely used in bakeries and dairies for forty years (Figure 6-7). These injector systems produce ice equivalent to ammonia-flooded performance with no more liquid refrigerant than used in direct expansion. However, practical pipe lengths limit this method to a 20-ton coil or multiple 20-ton coils in one tank. (160 ft. of 1 1/4" pipe per circuit or about 1.5 tons per circuit with R22 are practical limits.) Also, because a "limited" charge is used, no high-pressure receiver is needed and the condensing system must be dedicated to the ice builder.

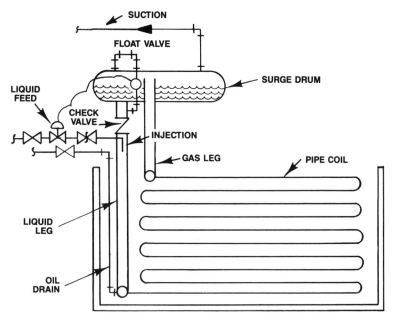

Figure 6-6.   Ammonia injection-aided circulation (for long circuits or overloaded storages)

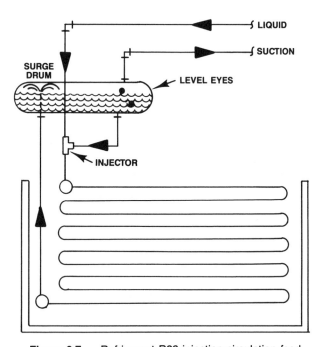

Figure 6-7.   Refrigerant R22 injection circulation feed

Multiple Low·Sides Injection Aided. When multiple ice builders are connected to a common condensing system, and a simple and efficient liquid injection system is desired, low-side pilot-operated float valves with distributors feeding individual injection points to each pipe coil can be used. (Figure 6-8). This results in a "top feed" circulating system that produces maximum ice harvests with minimum liquid charge. The condensing system requires a high-pressure receiver to accommodate fluctuations in the operating liquid charge caused by changes in head and suction pressures.

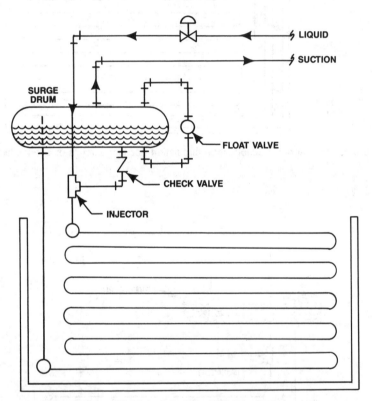

Figure 6-8.   Multiple low sides R22 injection circulation

Liquid Overfeed Systems

A liquid overfeed system is ideal for feeding refrigerant to the evaporator coils where a number of ice builders are connected to a common compression system. Three methods commonly used are:

Flash Gas Pump. Flash gas pumping (Figure 6-9) uses high-pressure refrigerant expansion to pump excess liquid through the coil. R-22 is well suited to this

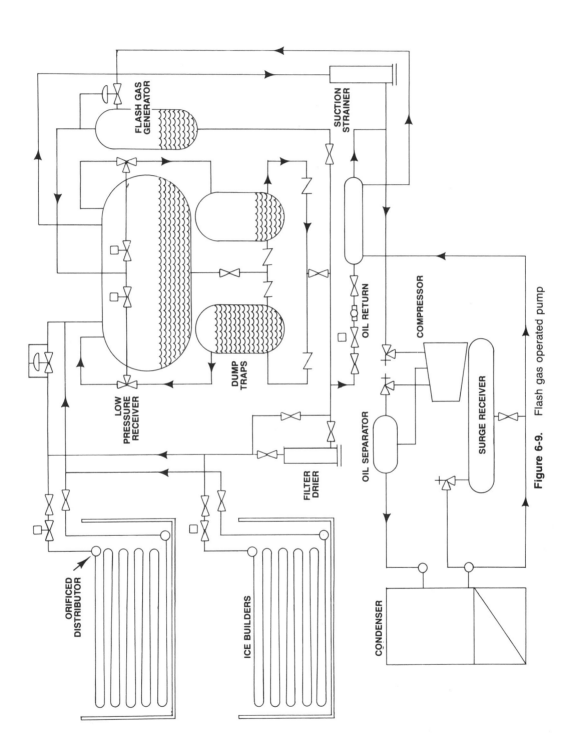

Figure 6-9.   Flash gas operated pump

Hot-Gas-Operated Pump. Hot gas pumping (Figure 6-10) can be economically applied by using a shell and coil suction trap for a direct expansion feed, and providing a gas-driven dump trap or pair of dump traps to handle the overfed liquid. The gas-operated pump only handles part of the flow, thus economizing on the gas use. Typically this use is about 2 percent of the compressor displacement for circulating 6 times the evaporating rate with ammonia at low pressure (10 to 5 psi) or for circulating 2 times the evaporating rate with R-22.

Gas pumping has an advantage over motor-driven pumping because neither a motor nor a mechanical pump is needed, thus reducing maintenance costs. Since the owning and operating costs of gas-driven pumps and motor-driven pumps are about the same, the choice of circulating equipment usually is dictated by the owners' or designers' preference.

Motor-Driven Pump. With motor-driven pumps (as with gas pumps), the purpose is to centralize the equipment to protect the compressor by mounting the accumulator between the ice tank and the compressor, to organize the liquid distribution, provide cold refrigerant feed by cooling in the accumulator to reduce evaporator pressure drop, and to increase the forced convection heat transfer by increasing the mass of liquid flow in the pipe coil by overfeeding and thus wetting more prime surface. Motor-driven pumps may be used at much higher overfeed rates than gas-driven pumps, using comparable power consumption, so they may show better performance on top-feed coils of short length. Bottom-fed circuits can be used with pumped circulation, but better heat transfer efficiency caused by more wetted surface while making ice does not occur, because the thermal conductivity of ice is the controlling resistance to heat flow. Also, top-fed circuits use much less refrigerant. There may be a slight advantage of bottom feed in the water chilling mode, after all the ice is melted, however, at the cost of considerably more refrigerant charge. A typical low-pressure pump receiver with oil return for R22 system is shown in Figure 6-11.

In summary, for refrigerant feed systems used in long pipe circuits, choose the simplest direct expansion method with a suction trap for dedicated compressor and ice tank applications. For ammonia systems using up to three ice tank coils on a common system, consider gravity circulation. For R22 systems with short pipe circuits using multiples up to three ice tank coils on a common system, consider injector-aided circulation.

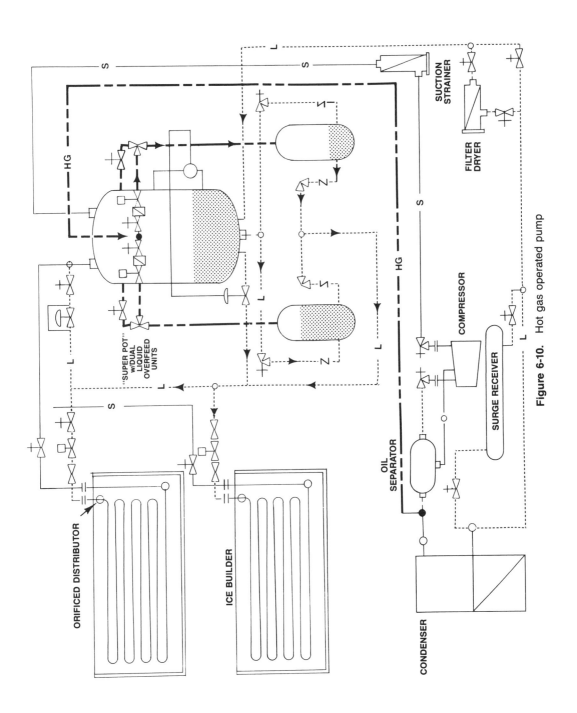

Figure 6-10. Hot gas operated pump

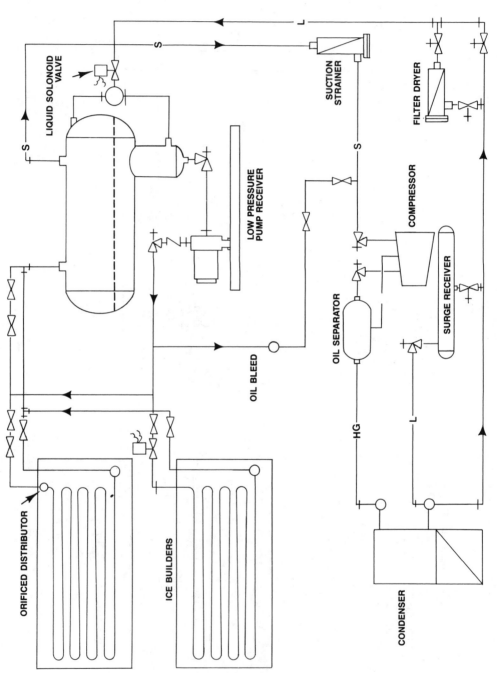

Figure 6-11. Low-pressure pump receiver with oil return for R22 system

For systems using more than three ice tank coils on a common system, consider a central accumulator with liquid overfeed system.

An operational requirement for all systems is the removal and return of oil from the evaporator or accumulator to the compressor. This is usually done manually with ammonia, but is easily automated with Refrigerant 22, as shown in Figures 6-2, 6-3, 6-9, and 6-11.

REFRIGERANT SELECTION

The refrigerants most commonly used in ice storage systems are ammonia and R-22 because they are relatively inexpensive, readily available, and have thermodynamic properties which result in minimal installed cost. Measured against these criteria, ammonia is preferable to R-22. However, because of occupancy rules in ANSI safety codes, ammonia is limited to industrial plants or separated facilities from personnel-occupied areas, so that the use of ammonia is limited to a few special situations for air conditioning using ice storage. Where ammonia or R-22 machinery rooms are used to house compressors, adequate ventilation fan volumes based on the mass of refrigerant in the system must be provided.

Table 6-1 provides a comparison of refrigerant performance per ton at standard fan conditions (5F evaporator and 86F condenser). On a coefficient of performance basis, R-502 is 9 percent less efficient than ammonia while R-22 and R-12 are within 2 percent. On a compressor displacement basis, R-12 needs 64 percent more machine volume than R-22 while R-22 and R-502 are within 5 percent of ammonia. The cost of ammonia and R22 compressors is practically the same. However, the evaporative condensers for ammonia must be 10 percent larger because the heat of rejection of ammonia is greater.

At 20F the density of ammonia liquid is 40.43 lbs/ft^3 and the density of R-22 is 81.60 lbs/ft^3. It therefore requires about twice the mass of R-22 to charge the ice builder coil. Since it costs 10 times more than ammonia, the refrigerant charge for R-22 can be twenty times the cost of ammonia.

The economics of refrigerant selection is also influenced by whether a bottom feed or top feed is employed. The condition inside the pipe coil with bottom feed is visualized as from 65 to 100 percent liquid phase. The condition inside the pipe coil with top feed is visualized as 25 to 40 percent liquid phase. Since there is no need for distribution orifices in bottom feed, and since there is more prime pipe surface wetted by refrigerant, there is some advantage in bottom feeding. The

Table 6-1

Comparison of Refrigerant Performance Per Ton

REFRIGERANT	COMPRESSION RATIO	REFRIGERANT CIRCULATED LB/MIN	LIQUID GPM	COMPRESSOR CFM	HP	COEFFICIENT PERFORMANCE
AMMONIA	4.94	0.422	0.085	3.44	0.989	4.76
REFRIGERANT 22	4.03	2.86	0.292	3.55	1.011	4.66
REFRIGERANT 12	4.08	4.00	0.371	5.83	1.002	4.70
REFRIGERANT 502	3.75	4.38	0.430	3.61	1.079	4.37

disadvantage of bottom feeding with R-22 is the increased cost of refrigerant, the increased potential for oil migration and compressor oil failure, and the increased loss potential in the event of a refrigerant leak. Also, an R-22 bottom-fed coil exhibits more pressure drop and thus has a higher evaporating temperature, causing less ice production than top-fed coils with long circuits. It thus is recommended that ammonia with thermal syphon be used when possible. A central system of liquid circulation should be used when there are more than three ice builders. Top feed should be used for R-22 systems regardless of the system used. R-502 and R-12 are not particularly well-suited for ice builder use.

RECEIVERS

Regardless of the refrigerant feed method chosen, the system will require a refrigerant receiver. One function of the receiver is to provide refrigerant storage space during system shutdown; another is to accommodate surging of the liquid volume between different operational modes and different load conditions. Regardless of receiver type, the vessel should be provided with a refrigerant relief valve (preferably dual relief valves manifolded on a changeover valve) and a drain valve.

High-Pressure Receivers. In both thermostatic expansion valve controlled systems or low side float controlled systems, a high pressure receiver is required between the condenser and evaporator to provide a constant liquid source to the TXV or float valve. A high pressure receiver is not needed with high side float valve systems, using low pressure receivers.

Low Pressure Receiver. The low pressure receiver is also called a suction trap or accumulator. Positioned between the compressor and evaporators, the vessel provides positive protection for the compressor against liquid slugging by separating liquid and vapor entering the vessel from wet return lines. It provides a liquid surge volume to absorb excess refrigerant charge and variations in level caused by changes in the liquid content of the pipe coil due to variations in head and suction pressures. The low pressure receiver also provides a liquid refrigerant ballast to feed the pipe coil during start up conditions before the liquid overfeed return has been established.

REFRIGERANT PIPING

When designing the connecting refrigerant piping of an ice storage system, careful attention must be given to pressure drop, compressor protection and oil return.

Pressure Drop

All piping (liquid discharge and suction) should be as small as possible to minimize cost and as large as possible to minimize operating penalties caused by pressure drop. Optimum sizing varies from system to system.

Refrigerant liquid piping from receiver to system presents minimum design problems. If pressure is reduced either because of line pressure loss or static head loss caused by an elevation change, some of the liquid will flash to the vapor state. Excessive flash gas results in reduced capacity and erratic operation of control valves. However, when control valves are below the receiver, static head gain will help to reduce flashing. If piping runs and elevations are such that excessive flash gas is generated, liquid subcooling should be employed. This may be done with a subcooling coil in the evaporative condenser or with a shell and coil suction-liquid heat exchanger in the system suction line, or with a shell and coil direct expansion subcooler, depending on the amount of subcooling that is necessary.

Refrigerant liquid drain lines from condenser to receiver should be designed to permit condenser liquid to freely drain from the condenser.

From the standpoint of pressure drop, sizing compressor discharge lines that carry hot gas to the condensers is straightforward. Because the piping involved transports high-pressure refrigerant, a relatively high-pressure drop per unit length is permissible. (Consult the ASHRAE Handbook of Fundamentals, Chapter 34, Pipe Sizing, for complete data on the selection of discharge line pipe size.)

The compressor suction line is the most critical aspect of refrigerant piping system design. Because it conveys a low-pressure vapor, the suction line is most sensitive to pressure drop. When a gravity-flooded or liquid recirculation feed system is used, the sole criterion for selecting the suction line size is pressure drop. However, when direct expansion feed is used with a halocarbon refrigerant, the suction line also is required to carry oil back to the compressor. The vapor must therefore have sufficient velocity to keep the oil entrained in it. This becomes particularly important in vertical pipe risers, which should be sized on the basis of the minimum velocity required to maintain oil entrainment. If these higher vapor velocities result in too great an overall pressure drop, the criteria for the horizontal runs must be reduced so as to effect a satisfactory system pressure drop. The ASHRAE Handbook Systems Volume, Chapter 26, System Practices for Halocarbon Refrigerants, and ASHRAE Handbook Fundamentals Volume, Chapter 34, Pipe Sizing contain complete data for properly sizing suction lines.

Compressor Protection

Reciprocating compressors and, to a lesser extent, rotary screw machines are sensitive to slugs of liquid refrigerant and/or oil. System piping must therefore use oil separators, traps, and suction-liquid heat exchangers. Much of this compressor protective equipment is related to effective oil return and is covered in more detail under the oil return section of this discussion.

In addition to protective equipment, the piping itself must be run to effect safe compressor operation. All suction piping should be pitched toward suction line traps or accumulators. All discharge piping should pitch toward the discharge line oil separator. Individual compressor connections should connect to suction and discharge headers above the center line. The riser to the evaporative condenser should rise above and connect into the top of the condenser inlet header. Finally, suction and liquid line strainers or catch-alls should be used as necessary to ensure flow of clean refrigerant to the compressors.

Oil Return

Reciprocating compressors require oil for lubrication. Rotary screw compressors require oil for lubrication, gas cooling and rotor sealing. Refrigerant contacts this oil in the compressor and, even with highly efficient discharge line oil separators, a small portion of the oil is carried around the refrigeration system, especially when halocarbon refrigerants (all miscible with oil) are employed. This oil tends to accumulate in the ice builder because it is the point of low pressure and temperature. Accordingly, the system should be designed to return the oil to the compressor.

Oil return systems require different treatment depending on the type of evaporator feed.

When thermal valve feed is used, oil return to a single compressor is accomplished by maintaining sufficient vapor velocity through the evaporator (ice builder coil) and the suction line.

When a liquid recirculating system with a refrigerant pump is used to feed the ice builder, the accumulator separates the liquid-vapor mixture returning from the ice builder. Oil entrained in the mixture stays in the accumulator and cannot return through the suction line. A tap off the liquid recirculating pump affords a simple and effective oil return. Pump discharge pressure is usually set at 20 to 25 psi above the suction pressure so there is ample pressure to carry the oil/refrigerant mixture to an expansion valve where it can be flashed to the lower suction line pressure. This pressure drop vaporizes the refrigerant liquid, and the resulting oil/vapor mixture can be piped into the screw compressor suction line. This arrangement is shown in Figure 6-12.

Oil recovery from ammonia systems is relatively simple since oil and ammonia are not miscible. Discharge oil separators are employed to recover any oil that is pumped out into the system by the compressors. The recovered oil is fed back to the crankcase of a reciprocating compressor through a float valve. (A screw compressor has an integral oil separator which feeds the recovered oil directly back to the compressor.) See ASHRAE Handbook Systems Volume, Chapter 27, System Practices for Ammonia, for complete recommendations.

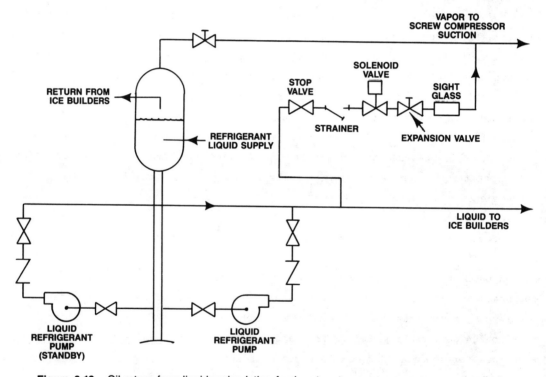

Figure 6-12. Oil return from liquid recirculation feed system to screw compressor suction line

Chapter 7

INSTRUMENTATION AND METERING

Effective cool storage system performance depends on the operator assistance designed into the system. For this reason the designer should consider the details of day-to-day operation, maintenance and system performance evaluation which will be handled by operating staff. Instrumentation is an important factor in this regard.

INSTRUMENTATION

The degree and type of instrumentation required depends on the size and complexity of the system, and the building operator's ability to use it. Also, if the designer is to periodically review performance, it may be desirable to have additional instrumentation to record and/or summarize performance data.

Indicating

Indicating instruments are installed to permit logs to be made, to observe that conditions are "proper" during periodic tours of the equipment, and to allow diagnosis of performance and operating problems.

Indicating-type instrumentation used for cool storage systems typically includes thermometers, pressure gauges, ammeters, pilot lights, flow indicators and hours operation counters. Where appropriate, these instruments should be adjustable to allow for calibration.

The ranges and accuracy of these devices should be selected with care. Thermometers should be capable of being read at 1F intervals or smaller. Pressure gauges should be capable of being read at 5-pound (or smaller) intervals. Extra thermometer wells and pressure gauge taps should be provided so that these instruments can be readily calibrated periodically. (A set of master calibrating instruments should be provided for this purpose.) Additional wells, taps and contacts should be provided for future use of data logging or computerized control.

Recording

The use of recording instruments should be considered for locations where there will be hourly and daily variations in measured parameters. They are also used to provide data to relate building cooling load and cooling storage system performance to outdoor temperatures. This type of information permits the operator to fine tune the system and improve his ability to predict cool storage requirements and thus the equipment which should be operated to provide both cooling storage and on-peak cooling. Where substantial variations in cooling load are likely to occur, or where uncertainties exist as to what these loads will be, hard-copy recorded data will prove to be invaluable for proper plant operation.

Recording instruments are available with daily, weekly and monthly charts or tapes. The most common types use circular or strip charts. Circular charts usually are changed daily or weekly; strip charts, weekly or monthly. Some of the newer recorders provide digital output, but most building operators prefer graphic recorders so they can readily see if proper operation is occurring. As with indicating instruments, it is important to specify the range and accuracy of recording devices, and to have a calibration capability.

In those cases where automatic data monitoring and recording are required, magnetic tape recorders typically are employed for the purpose. These devices record several information channels side-by-side on the tape, with one channel being expressly reserved for recording time pulses generated by an internal clock. Other information commonly recorded includes chilled water Btu consumption and the electrical consumption of the compressor/condenser unit, pumps and other equipment. Such data are recorded continuously, usually at the rate of seven inches (of tape) per hour. In addition to permanently installed recording instruments, portable recorders often are employed to examine the performance of individual pieces of equipment.

ENERGY MANAGEMENT AND CONTROL SYSTEMS

In those buildings where central energy management and control systems are utilized, many indicating and recording functions can be performed by the central computer system. An analysis should be performed first, however, because -- particularly in smaller buildings -- it often is less expensive to install self-contained instruments at the locations desired, rather than collecting information at the central computer system. In larger buildings or in buildings where a large number of recorders are required, centralizing likely will be less expensive and more convenient. Nonetheless, even if instruments are provided in connection with a central computerized system, it still is desirable to have indicating instruments located at the points throughout the system where these readings are necessary.

Note that many small, inexpensive, microprocessor-based energy management and control systems can also log performance data, including hourly electric use and a variety of temperatures. Some systems can automatically call a remote computer terminal over the telephone and print out data periodically, as requested. Calls can also be initiated to the unit from any remote dial-up tape terminal, to obtain the type of data and numerous other reports. With such a system, data is available both to the operator and the designer.

METERING

Utility companies utilize both indicating and recording electric demand and kilowatt hour meters, with recording meters typically being used in larger buildings.

It is most desirable to have recording-type electric meters on buildings which employ cool storage, so that cool storage system performance can be readily ascertained. If the utility would ordinarily install an indicating-type demand meter, a recording-type meter can be requested; a small monthly premium may be imposed.

If recording-type meter is not available from the utility, one can be installed on the customer side of the utility meter. Recording electric meters usually are available for less than $1,000, including 32-day circular chart or strip chart recorders, magnetic tape cassette recorders and electronic microprocessor-type meters. Circular or strip charts are easily read and can be changed by building staff. It is necessary to employ an outside service to obtain hard copy data from magnetic tape cassette recorders.

It also is desirable to install submeters throughout the building, especially if significant demand costs are involved. Recording types can be installed for the cool storage system and the remaining building load. The information derived allows the operator to better integrate the operation of the cool storage system with electric loads over which there is little or no control, to minimize electric demand. Additional metering should be considered for electric heating and/or electric domestic water heating loads and each cooling compressor, substation, feeder, motor control center and major item of equipment. (By installing an electric meter on a motor control center in conjunction with hours operation counters on each starter, it is possible to allocate the electricity consumed at the motor control center to each motor with a reasonable degree of precision.)

When submeters are incorporated in the original construction of the building and its equipment, their cost is low compared with that associated with installing them at a later date. However, even if meters are not installed at the time the building is built, it will be worthwhile to make provisions for their installation at some future date, by providing space for the installation of potential transformers and current transformers, as well as installing meter sockets or meter pans.

MONITORING

It generally is desirable to monitor cool storage system performance closely, to permit adjustments which will help assure continuous optimization of system performance. Monitoring also permits rapid indication of problems, in turn supporting an expeditious remedial response. Monitoring systems can range from relatively simple and inexpensive to relatively complex and costly. The extent of investment appropriate usually depends upon the size of the system involved, in that the larger the system, the more savings to be derived from continous fine-tuning.

The most advanced systems perform both monitoring and data logging. These can be integrated with energy management and control systems (EMCS) which can be programmed to take optimizing control actions automatically.

Typical monitoring devices employed for cool storage systems include temperature sensors, flow meters, Btu meters, air flow indicators, and watt-hour meters as follows:

Temperature Sensors

A platinum resistance temperature detector (RTD) with a range of 0-150F and accuracy of $\pm$ 0.1% is most commonly employed. When used to monitor chilled water temperature, an RTD is mounted in a stainless steel thermal well. When the RTD is employed to measure outdoor wet-bulb temperature, an automatic welling system is used.

Flow Meters

Fluid flow measurements are made using turbine flow meters equipped with pulse generators and ancillary devices. Since output frequency of the turbine flow meter is proportional to flow rate, every pulse from the turbine flow meter is equivalent to a known volume of fluid that has passed through it. Electronic counters totalize these pulses to identify total volumetric flow.

Btu Meters: The total cooling load is monitored by Btu meters used with temperature sensors and turbine flow meters. The temperature sensors are installed at the cool storage unit's supply and return chilled water headers. The turbine flow meter is installed in the chilled water supply line to the cooling coils. A pulsing unit attached to the flow meter sends a flow rate signal to the Btu meter. The Btu meter's logic system multiplies the flow rate by the water temperature differential and integrates the product over time to yield the amount of Btu's of cooling provided by the cool storage unit. The Btu meter generates pulsed signals which are transmitted to a magnetic tape recorder.

Air Flow Indicators

Air flow rates are measured by devices which use arrays of pitot and static pressure tubes to produce an average pressure difference. The difference is sensed by a diaphragm-type transducer.

Watt-Hour Meters

Electrical energy use is measured by watt-hour meters equipped with pulse generators.

Chapter 8

SYSTEM PROCUREMENT

Cool storage systems share many of the same components as a conventional system, however, there are many variances. While any experienced mechanical contractor should be able to deal with a cool storage system with relative ease, some may be fearful because of lack of experience. In any event, it should be recognized that cool storage systems and conventional systems are not the same. A successful installation is most likely to result if both the designer and contractor stress quality, and apply the type of effort which each is capable of, but which various emergencies (typically associated with time and money) often preclude. The following considerations should be observed.

CONTRACTING METHODOLOGY

Two general approaches to contractor selection are available: bidding and negotiation. Bidding encourages all interested firms to submit as low a price as possible, but seldom is the price actually paid as low as the winning bid. Typically, change orders occur and these lead to price increases, delays, and other problems. Accordingly, if it is decided to employ bidding, it is suggested strongly that the award not necessarily go to the low bidder. Instead, all firms should be prequalified, and a prebid conference should be held. Contractors also should be required to include in their bids appropriate price schedules to cover contingencies and extras.

A negotiated approach is considered superior to bidding because it permits selection of a contractor based principally on quality factors. It does not result in a contractor receiving a "blank check," but it does give the latitude required to perform the work well, for a reasonable, usually "not to exceed" price.

It may be appropriate to rely on a non-local contractor with experience with cool storage systems but, in such a case, it may be appropriate to have a local contractor involved as well, to help assure compliance with local codes and timely interaction with various governmental and other individuals. It may also be appropriate to call for a joint venture-type submission (in the case either of bidding or negotiation) to help assure that the contracting team has familiarity both with cool storage technology and local conditions.

BIDDING DOCUMENTS

The amount of detail associated with plans and specifications for a cool storage system should greatly exceed that associated with conventional systems, because most interested contractors will not be in a position to make confident assumptions, due to their likely lack of familiarity with the system being specified.

Table 8-1 comprises a suggested list of items included in a set of drawings. Table 8-2 comprises a check list, indicating some of the most significant details which should be included in specifications. Close coordination between drawings and specifications is mandatory. It is also essential to provide a high level of coordination with other design professionals involved on the project to help minimize the need for changes during construction. In addition, it also is essential to afford continuous on-site construction monitoring, to help assure adherence to plans and specifications, as well as prompt response to any questions which may arise or changes which may be needed. In fact, several major studies have suggested that inadequate communication and coordination, and inadequate on-site monitoring have been principal causes of many of the problems which have resulted in so much construction industry litigation and low productivity (15). A report developed by an oversight subcommittee of the Committee on Science & Technology of the U.S. House of Representatives states that better coordination and more on-site monitoring could be significant in reducing the frequency of structural failures in the United States.

Table 8-1

List of Items Included in Drawings

- Legend and Symbol List

- Floor Plan

    - Location of Cool Storage Equipment
    - Location of Controls
    - Related Mechanical/Electrical Modifications (in case of existing installations)

- System Schematic Diagram

- Flow Diagram (water, brine and refrigerant system)

- Control Diagram

- Storage Tank Installation Devices

- Instrumentation and Metering Devices

Table 8-2

Specification Check List

GENERAL	ICE/CHILLED WATER STORAGE UNIT

WORK DESCRIPTION
SUPERVISION
PERMITS & REGULATIONS
GENERAL NOTES
EXISTING SYSTEM & CONTROL
   COMPRESSORS/CHILLERS
   AIR DISTRIBUTION
   HEAT EJECTING DEVICES
   CONTROLS
MODIFICATION
UTILITIES
	GENERATING CAPACITY (TON-HOURS)
	STORAGE CAPACITY
	WATER/BRINE/REFRIGERANT
	   FLOW RATE, MAXIMUM
	   TEMPERATURE, IN
	   TEMPERATURE, OUT
	   RESISTANCE TO FLOW
	TANK CAPACITY (CU. FT.)
	STORAGE TANK
	   CONSTRUCTION
	   INSULATION
	   BAFFLING
	   AGITATION
	   CORROSION PROTECTION
	   EVAPORATOR CONSTRUCTION
	CONTROL
	LOCATION

COMPRESSOR/CHILLERS	

QUANTITY
TYPE
   RECIPROCATING
   SCREW
   CENTRIFUGAL
REFRIGERANT TYPE
TEMPERATURE
   SUCTION
   CONDENSING
CAPACITY (TONS)
LIQUID
   TYPE: WATER, BRINE
   SOLUTION CONCENTRATION
   DENSITY
   VISCOSITY, KINETIC
LIQUID TEMPERATURE
   ENTERING
   LEAVING
CAPACITY CONTROL
LOCATION
ELECTRICAL CHARACTERISTICS
   HORSEPOWER
   VOLTAGE
   PHASE
   LOCKED ROTOR AMPS
   FULL LOAD AMPS
   MOTOR STARTER
	CHILLED WATER/BRINE DISTRIBUTION SYSTEM
	
	TYPES
	   CONSTANT WATER FLOW
	   VARIABLE WATER FLOW
	PIPING
	   MATERIAL
	   SYSTEM PRESSURE
	BRINE
	   TYPE
	   PERCENT SOLUTION
	FLOW RATE (GPM)
	SUPPLY TEMPERATURE (F)
	RETURN TEMPERATURE (F)
	CONTROL VALVES
	PIPING INSULATION
	SYSTEM BALANCING
	AIR SEPARATOR
	   LIQUID FLOW RATE, MAXIMUM
	   PRESSURE DROP, MAXIMUM
	EXPANSION TANK
	   EXPANSION CAPACITY
	   TYPE
	      MECHANICALLY PRESSURIZED
	      SYSTEM PRESSURIZED
	   PRESSURE
	      MINIMUM
	      MAXIMUM

HEAT EJECTION/RECOVERY DEVICES	CONDENSER WATER SYSTEM

CAPACITY
TYPE
   AIR COOLED
   WATER COOLED/COOLING TOWER
   EVAPORATOR
   HEAT RECOVERY/HEAT EXCHANGER
CONDENSER EQUIPMENT
   CONDENSING TEMPERATURE
   AMBIENT WET BULB TEMPERATURE
   FOULING FACTORS
   FAN POWER (KW)
   CONDENSER WATER TEMPERATURE
      ENTERING
      LEAVING
LIQUID RECEIVER (CU. FT.)
CAPACITY CONTROL
HEAT EXCHANGER
   TYPE
   FLOW RATE
   TEMPERATURE
      ENTERING
      LEAVING
LOCATION
ELECTRICAL REQUIREMENTS
   HORSEPOWER
   VOLTAGE
   PHASE
	PIPING
	FLOW RATE (GPM)
	SUPPLY TEMPERATURE
	RETURN TEMPERATURE
	CONTROL VALVES
	
	REFRIGERANT PIPING
	
	MATERIALS
	PRESSURE DROP
	OIL RETURN
	
	PUMP(S)
	
	QUANTITY
	TYPE
	   CENTRIFUGAL
	   TURBINE
	   OTHER
	LIQUID FLOW RATE
	PUMPING HEAD
	LIQUID FLOW RATE CONTROL
	   MULTIPLE PUMPS
	   VARIABLE SPEED

Table 8-2 (cont'd)

Specification Check List

PUMP(S) (cont.)	INSTRUMENTATION, METERS & RECORDERS

ELECTRICAL CHARACTERISTICS
   POWER EACH PUMP
   POTENTIAL
   PHASE
   FULL LOAD
   LOCKED ROTOR
MOTOR STARTER
   ACROSS-THE-LINE
   INCREMENT
   VARIABLE SPEED
	TEMPERATURE SENSORS
	HYGROMETER RECORDER
	ICE THICKNESS INDICATORS
	FLOW METERS
	BTU METERS
	KILOWATT-HOUR METERS
	RUNNING TIME METERS
	STRIP CHART RECORDERS
	MAGNETIC TAPE RECORDERS

PUMP(S) (cont.)

ELECTRICAL CHARACTERISTICS
  POWER EACH PUMP
  POTENTIAL
  PHASE
FULL LOAD
LOCKED ROTOR
MOTOR STARTER
  ACROSS-THE-LINE
  INCREMENT
  VARIABLE SPEED

AIR DISTRIBUTION SYSTEM

AIR HANDLING UNIT
  AIR FLOW RATE
  TOTAL STATIC PRESSURE
  BLADE CONFIGURATION
  ELECTRICAL REQUIREMENTS
VARIABLE AIR VOLUME CONTROL
  MULTIPLE FANS
  INLET (VORTEX) DAMPER
  OUTLET DAMPER
  BY-PASS DAMPER
  MULTIPLE SPEED
  VARIABLE SPEED
ECONOMIZER CONTROL
VOLUME CONTROL
  SHEAVE ADJUSTMENT
COOLING COIL
  CAPACITY
  AIR & WATER FLOW RATE
  AIR & WATER TEMPERATURE (ENT. & LEAV.)
  RESISTANCE TO FLOW
FILTER
  TYPE
  AIR FLOW RATE
  RESISTANCE
  EFFICIENCY

CONTROLS

TYPE
  ELECTRONIC
  ELECTRIC
  PNEUMATIC
MOTORS AND MOTOR CONTROLS
PIPING AND CONTROL WIRING
CONTROL EQUIPMENT
  THERMOSTATS
  HUMIDISTATS
  VALVES, 2-WAY, 3-WAY
  AUTOMATIC DAMPERS
  VARIABLE AIR VOLUME
  SAFETY CONTROLS
  DIFFUSER CONTROLS
  DAMPERS
LOCAL CONTROL PANELS
ENERGY MANAGEMENT & CONTROL SYSTEMS

INSTRUMENTATION, METERS & RECORDERS

TEMPERATURE SENSORS
HYGROMETER RECORDER
ICE THICKNESS INDICATORS
FLOW METERS
BTU METERS
KILOWATT-HOUR METERS
RUNNING TIME METERS
STRIP CHART RECORDERS
MAGNETIC TAPE RECORDERS

MAINTENANCE

COMPRESSORS/CHILLERS
CONDENSERS/COOLING TOWERS
HEAT EXCHANGERS
COOL STORAGE UNIT
  CORROSION INHIBITION
  REPAINTING
DISTRIBUTION SYSTEMS
  AIR
  WATER
  BRINE
  REFRIGERANT
CONTROLS
  CALIBRATION
  ADJUSTMENT
INSTRUMENTATION, METERS & RECORDERS

ACCEPTANCE TESTING, START-UP, DEBUGGING AND TRAINING

Specifications should require the mechanical engineer of record and the contractor to perform acceptance testing, system start-up, system debugging and provide operator training. These needs are discussed in Section 9.

Chapter 9

OPERATION AND MANAGEMENT

As with any type of cooling system, the care and attention provided in operation
and maintenance have a significant influence on the system's performance and
reliability, as well as its operating and maintenance costs and longevity.
Generally speaking, most operating and maintenance (O&M) persons will be more
familiar with chilled water cool storage systems than ice storage systems.
However, in either case, long-time habits may need correcting. Most O&M personnel
are used to operate conventional systems in response to cooling demand.
Significantly different concerns are present when cool storage is employed.
Applying "standard operating procedures" will result in unacceptable system
performance.

As indicated below, the system designer can provide services and assistance which
will help assure optimum results.

OPERATION AND MAINTENANCE STAFFING

The technical expertise required of those operating and maintaining will depend
upon the complexity of the system installed. In some cases, knowledge of
conventional systems augmented through training relative to the specific cool
storage system involved will suffice. While many smaller conventionally cooled
buildings do not have any full-time people involved in cooling system operation,
one person with at least part-time responsibility for cool storage system operation
and maintenance will be needed regardless of building size. And when only one
person is involved, it will be important to have one or more maintenance and
service contracts with an outside contractor experienced with cool storage. These
factors should be considered not only in economic analysis, but also in system
design.

The building owner should be advised about the number of people and capabilities
required. He should be advised about the supervision necessary for the cool
storage system, and the building manager should be given a simplified description
of the system. The building manager should also be told what to expect in the way

of performance and operation, as well as the type of information and reports which should be provided to them by the operating engineers and outside contractors.

OPERATING AND MAINTENANCE INSTRUCTIONS

The installing contractor probably will provide a project manual relative to the cool storage system, and equipment manufacturers ordinarily provide information about maintenance and spare parts. Neither will be sufficient to provide the type of guidance required for proper operation and maintenance and, for this reason the designer should provide comprehensive operating and maintenance instructions that can be readily understood both by staff and any outside contractors. Additional details will be desirable when staff and/or local contractors are not familiar with cool storage systems. (The designer should also identify what spare parts unique to the cool storage system should be furnished with the original construction, especially in areas where these parts may not be readily available.)

The designer's operating instructions should include clear, readily understandable descriptions of his concepts and the manner in which the cool storage system should operate. They should emphasize the unique features of the cool storage system, so that the operator can pay particular attention to them. It probably will be necessary to modify the original operating instructions based on experience gained during the first year of operation.

Operation

Operating instructions should be provided both for initial annual start-up and for day-to-day operations. Start-up instruction should pay particular attention to the manner in which each element of the cool storage system performs and how they perform together.

Day-to-day operating instructions should address selection of equipment to be used and the unique aspects of the system. These aspects include the extent to which automatic controls are intended to function and the degree of manual intervention. Operating instructions should also include information on utilizing data obtained from instrumentation, relative to the quantity of cooling energy stored and the quantity removed.

The quantity of cooling energy to be stored on any given day will depend on the cooling load anticipated for the next day, based on weather forecasts and experience. Generally speaking, it is best to err on the safe side, and store more cooling energy than would otherwise be anticipated.

Fully charging the cooling storage system prior to the start of each day's operation is a conservative operating approach. It is acceptable as long as the losses are not substantial. Nonetheless, instrumentation should be used to measure these losses and their cost should be determined and weighed against the cost of manual intervention and/or automatic controls. If the cost of losses associated with conservative operation is significant, it will be necessary to establish operating instructions and/or to install automatic controls which can charge only to the extent necessary, with a safety factor. Typically, this will involve preselecting the number of compressors to be operated and/or the number of hours they are to be operated.

Where partial cool storage is provided by design, instructions will be necessary to determine the number of compressors to be run during daytime hours, and the conditions which affect compressor start-up and shut-down times.

For buildings with a consistent cooling load, cool storage system operation will tend to be relatively routine. More operator attention and/or more control automation is necessary for buildings with variable cooling loads.

In the case of highly variable hour-to-hour cooling loads, especially when there are relatively extreme peak hourly loads, operation of the cool storage system can become complex. If storage capacity is marginal compared with cooling loads, proper operation becomes critical. In such cases, it will be necessary to determine the consequences of any inability to meet cooling load, such as loss of humidity or temperature control, situations which may occur only on humid days or days when unique cooling loads occur. (Closing outdoor air dampers (where possible) is a potential means for load reduction during these unique situations.) If it becomes necessary to operate more cooling compressors than anticipated during on-peak hours in a partial storage system, operating cost consequences will have to be weighed against the consequences of being unable to meet the load.

Operating instructions should anticipate off-normal situations and consider load-shedding in noncritical areas and/or pre-cooling to avoid on-peak compressor operation.

Methods of accomplishing seasonal shut-down should also be included in operating instructions. On occasion some cool storage equipment will break down, making it necessary to operate the system in other than the originally intended method. Operating instructions should anticipate as many of these eventualities as possible, and provide instructions on how to maintain comfort conditions while repairs are accomplished. (Since the cost penalty associated with breakdowns can be substantial, estimates of these costs can be used to evaluate the appropriateness of standby equipment. Where the cooling loads served are not critical, it will be necessary to estimate whether diminished cooling or no cooling at all can be tolerated while repairs are performed.)

Maintenance

Maintenance instructions should include the special requirements of a cool storage system in addition to the routine maintenance associated with any conventional cooling system. Typical routine maintenance associated with cool storage includes calibration of instruments and controls, inspections and other checks for losses and leakage, and checks on ice and water level in tank and water treatment requirements and the condition of the cool storage equipment. Note, storage equipment maintenance requirements are significantly influenced by the materials used in its fabrication as well as its location and the surrounding environment. Typically storage tanks are fabricated from steel, polyethelyne, fiberglass or concrete. Of all these types, the steel tanks require the most maintenance. Steel storage tanks should be checked periodically for corrosion and the condition of their insulation. These tanks are painted inside with a rust-resistant primer, and repainting is necessary because of weathering and general wear.

The water/glycol solution used in certain types of ice storage systems must be checked periodically in order to ensure that proper inhibition is maintained.

Further, instructions regarding handling of refrigerants should be provided. Special precautions are necessary when handling ammonia, because of its high toxicity. Ammonia is highly irritant and the odor of ammonia gives instant warning of a leak. Ammonia vapor will readily dissolve in any moisture to form a caustic solution, which is corrosive to copper and most copper alloys. Special bronzes have been developed which are not corroded by ammonia containing water vapor. Aside from these special bronzes, no copper, brass, or bronze should be used in pipes, fittings, and bearings in an ammonia system.

The maintenance instructions should also include work to be performed during the winter, including any teardown of equipment and cleaning requirements of the storage equipment.

When it becomes necessary to drain water from a cool storage system, careful attention should be paid to water treatment subsequent to refilling. (Means for performing the occasional draining and refilling that will be necessary should be provided.) Large, possibly multiple drains, and one or more refilling points can be used for the purpose. Make-up water should be metered to determine if there are water losses from the system, and the magnitude of those losses.

Table 9-1 identifies certain major HVAC components, the types of preventative maintenance they require, and the frequency with which that maintenance should be performed.

Operating Logs: Operating logs must be maintained. They are used to provide historical information on cool storage system performance and to permit the operator to gauge actual performance against predictions. This data also is used to prepare periodic performance summaries. Samples of operating logs for ice and chilled water storage are shown in Tables 9-2 and 9-3. Note that some manufacturers provide model logs developed specifically for their equipment.

The level of detail in the logs is influenced by the extent of recording instruments; the more recordings made, the less detail required, in that charts obtained from the recorders can be used with the logs. Space should be provided for computed daily consumption for each meter read, so that it ultimately becomes possible to predict the performance of individual pieces of equipment by averaging historical data contained on the logs and charts. By comparing predicted performance to actual performance each day, it will be possible to identify those operating practices which affect performance, and thus how performance can be optimized.

The designer should prepare initial drafts of the operating logs in conjunction with the chief operating engineer. After some operating experience has been gained, the content and format of the logs should be finalized and printed.

FINE TUNING

The designer should be present during initial start-up and shakedown to observe system performance and to answer any questions that may be posed by the various contractors and the operating engineers. By observing operation, the designer will be able to refine the operating instructions and advise the operating engineer regarding preferred methods of operation.

Table 9-1

Suggested Frequency for Maintenance Schedule

D = Daily
W = Weekly
M = Monthly
S = Semi-Yearly
Y = Yearly
X = As Needed

	ALIGNMENT	LUBRICATION	OVERHEATING	BELT TENSION AND ALIGNMENT	BLOW OUT DRIP POCKETS AND ELIMINATORS	CLEAN	AIR TAKE	DUST FROM	FILTERS	OIL CHAMBERS	SCREENS	SPRAYS	STRAINERS AND TANKS	CONTROLS	SETTING AND CALIBRATION	PROTECTIVE DEVICES	DRAINAGE	EFFECTIVE OPERATION	FREEZE PROTECTION	INSPECT GENERALLY	LEAK DETECTION AND REPAIR	OIL LEVEL	OIL PRESSURES	OPERATING PRESSURES	PUMP DOWN	REFRIGERANT LEVEL	ROTATION	SPRAY EROSION	STANDBY	SEALS	Ph VALUE OF WATER
AIR COMPRESSOR	X	Y		M	W		Y		S	Y	W	M	W	Y						Y		M									
CIRCUIT BREAKER OR STARTER		Y												Y	Y					Y	Y										
COILS								S									S		S	Y	Y			S							
CONDENSERS																	S		S	Y	Y				D						W
CONDITIONED SPACE															W					W											
CONTROLS					W	Y									W	M				S											
COOLING TOWERS		M	S									S	X	S	W			M	S	S	Y	S		M					S	S	W
DAMPERS	S	Y												S				M		Y											
DISTRIBUTION SYSTEM						X	X													Y											
DRIVES	M			M																Y											
EVAPORATIVE CONDENSER		S	S									W	S	W					S	S	Y			W					M		
FANS	X	M	S	M				Y												X	M						Y				
FILTERS							Y	Y	S				X	M			M			Y	W										
FREEZE PROTECTION														S			S		S	S											
MOTORS		M	S	M				S		Y			Y							Y	M						Y				
OPERATING CONDITIONS																		D													
OPERATING SCHEDULE																		S													
PUMPS	Y	M	S							Y	X		X						S	Y	M								S	Y	
REFRIGERANT PIPING																					M			S							
REFRIGERATION COMPRESSOR	S	M	M	M				X						M						Y			D	D	D	S		W	S	W	
STORAGE EQUIPMENT						M								S	W		S			W	M										
WATER CONDITIONING															W		W			W											
WATER PIPING					S								W				S		S	Y									S		W

Table 9-2

Equipment Operating Log
Ice Storage System

DATE _____
TIME OF DAY _____
EQUIPMENT IDENTIFICATION NUMBER, IF ANY _____

WATER CHILLING PACKAGE, RECIPROCATING TYPE
WATER COOLER
  WATER
      TEMPERATURE-   IN   DEG. F.   _____   _____
                            OUT   DEG. F.
      PRESSURE-   IN   PSIG
                            OUT   PSIG
    FLOW RATE (1)   GPM
  REFRIGERANT
    TEMPERATURE, ACTUAL SUCTION   DEG. F.
CONDENSER
  WATER
      TEMPERATURE-   IN   DEG. F.
                            OUT   DEG. F.
      PRESSURE-   IN   PSIG
                            OUT   PSIG
    FLOW RATE (2)   GPM
COMPRESSOR
  BEARING OIL PRESSURE   PSIG
  REFRIGERANT
    PRESSURE   DISCHARGE   PSIG
                          SUCTION   PSIG
  UNLOADED   %

HEAT EJECTION SYSTEM
CONDENSER WATER PUMP(S) NUMBER
  PRESSURE-   IN   PSIG
                            OUT   PSIG
  FLOW RATE (3)   GPM
COOLING TOWER
  CELLS IN OPERATION   Y OR N
  AMBIENT AIR-TEMPERATURE   DB/WB
           BAROMETRIC PRESSURE   IN.HG
  CONDENSER WATER
      TEMPERATURE-   IN   DEG. F.
                            OUT   DEG. F.
    WATER FLOW RATE (3)   GPM
  PAN, CELL
      WATER LEVEL, NORMAL   Y OR N
      TEMPERATURE (WINTER IF FILLED)   DEG. F.
    FANS OPERATING   Y OR N
WATER FILTERING SYSTEM
  OPERATING   Y OR N
  FILTER MEDIA
      PRESSURE-   IN   PSIG
                            OUT   PSIG
  BACK-WASHED   Y OR N
  PUMP
      PRESSURE-   IN   PSIG
                            OUT   PSIG
    WATER FLOW RATE (3)   GPM
AIR COOLED CONDENSER
  UNITS OPERATING
  AMBIENT AIR
    TEMPERATURE   DB/WB
    BAROMETRIC PRESSURE   IN.HG.
  FANS OPERATING   QUAN.

Table 9-2 (cont'd)

Equipment Operating Log
Ice Storage System

BRINE CIRCULATING SYSTEM				
PUMP				
PRESSURE-	IN	PSIG	_____	_____
	OUT	PSIG	_____	_____
BRINE FLOW RATE (4)				
BRINE FLOW TO USER EQUIPMENT				
TOTAL RATE		GPM	_____	_____
TEMPERATURE-	TO	DEG. F.	_____	_____
	FROM	DEG. F.	_____	_____
FLOW RATE		GPM	_____	_____
TANK				
ICE FORMING				
STARTS, TIME		HR/MIN	_____	_____
TERMINATES, TIME		HR/MIN	_____	_____
BRINE				
FLOW RATE (4)		GPM	_____	_____
TEMPERATURE	ENTER	DEG. F.	_____	_____
	EXITS	DEG. F.	_____	_____
ENERGY WITHDRAWAL				
STARTS, TIME		HR/MIN	_____	_____
TERMINATES, TIME		HR/MIN	_____	_____
BRINE TO USING EQUIPMENT				
FLOW RATE (5)		GPM	_____	_____
TEMPERATURE	TO	DEG. F.	_____	_____
	FROM	DEG. F.	_____	_____

NOTES:
(1) From flow meter or derive from cooler pressure drop/flow curve using pressure differential converted to feet.
(2) From flow meter or derive from condenser water pump (if one for each condenser) flow/head curve using pump inlet and outlet pressure differential converted to feet.
(3) From flow meter or derive from pump flow/head curve using inlet and outlet pressure differential converted to feet.
(4) From flow meter or derived from brine pump gpm/head curve using pump inlet and outlet differential pressure converted to feet.
(5) Obtain from flow meter or may be the same as pump flow.

Table 9-3

Equipment Operating Log
Chilled Water Storage System

DATE _____
TIME OF DAY _____
EQUIPMENT IDENTIFICATION NUMBER, IF ANY _____

WATER CHILLING PACKAGE, CENTRIFUGAL TYPE
WATER COOLER
 WATER
  TEMPERATURE-   IN   DEG. F.   _____ _____
                       OUT   DEG. F.   _____ _____
  PRESSURE-   IN   PSIG   _____ _____
                       OUT   PSIG   _____ _____
  FLOW RATE (1)   GPM   _____ _____
 REFRIGERANT
  PRESSURE   IN.HG   _____ _____
  TEMPERATURE, ACTUAL SUCTION   DEG. F.   _____ _____
CONDENSER
 WATER
  TEMPERATURE-   IN   DEG. F.   _____ _____
                       OUT   DEG. F.   _____ _____
  PRESSURE-   IN   PSIG   _____ _____
                       OUT   PSIG   _____ _____
  FLOW RATE (2)   GPM   _____ _____
 REFRIGERANT
  PRESSURE   IN.HG.   _____ _____
  TEMPERATURE- ACTUAL DISCHARGE DEG. F.   _____ _____
    HIGH PRESSURE, LIQUID   DEG. F.   _____ _____
COMPRESSOR
 BEARING OIL PRESSURE   PSIG   _____ _____
 PREROTATION VANE POSITION   % OF OPEN   _____ _____
OIL RETURN SYSTEM
 OPERATING   Y OR N   _____ _____
NONCONDENSIBLES PURGE SYSTEM
 CHAMBER PRESSURE   PSIG   _____ _____
 EXCESS PURGES?   Y OR N   _____ _____
MOTOR
 CURRENT- INPUT   AMPERES   _____ _____
 POTENTIAL,   VOLTS   _____ _____

WATER CHILLING PACKAGE, RECIPROCATING TYPE
WATER COOLER
 WATER
  TEMPERATURE-   IN   DEG. F.   _____ _____
                       OUT   DEG. F.   _____ _____
  PRESSURE-   IN   PSIG   _____ _____
                       OUT   PSIG   _____ _____
  FLOW RATE (1)   GPM   _____ _____
 REFRIGERANT
  TEMPERATURE, ACTUAL SUCTION   DEG. F.   _____ _____
CONDENSER
 WATER
  TEMPERATURE-   IN   DEG. F.   _____ _____
                       OUT   DEG. F.   _____ _____
  PRESSURE-   IN   PSIG   _____ _____
                       OUT   PSIG   _____ _____
  FLOW RATE (2)   GPM   _____ _____
COMPRESSOR
 BEARING OIL PRESSURE   PSIG   _____ _____
 PRESSURE   DISCHARGE PSIG   _____ _____
                     SUCTION   PSIG   _____ _____
 UNLOADED   %   _____ _____

HEAT EJECTION SYSTEM
CONDENSER WATER PUMP(S) NUMBER
 PRESSURE-   IN   PSIG   _____ _____
                       OUT   PSIG   _____ _____
 FLOW RATE (3)   GPM   _____ _____
COOLING TOWER
 CELLS IN OPERATION   Y OR N   _____ _____
 AMBIENT AIR-TEMPERATURE   DB/WB   _____ _____
    BAROMETRIC PRESSURE   IN.HG   _____ _____
 CONDENSER WATER
  TEMPERATURE-   IN   DEG. F.   _____ _____
                       OUT   DEG. F.   _____ _____

Table 9-3 (cont'd)

Equipment Operating Log
Chilled Water Storage System

WATER FLOW RATE (3)		GPM		
PAN, CELL				
WATER LEVEL, NORMAL		Y OR N		
TEMPERATURE (WINTER IF FILLED)		DEG. F.		
FANS OPERATING		Y OR N		
WATER FILTERING SYSTEM				
OPERATING		Y OR N		
FILTER MEDIA				
PRESSURE-	IN	PSIG		
	OUT	PSIG		
BACK-WASHED		Y OR N		
PUMP				
PRESSURE-	IN	PSIG		
	OUT	PSIG		
WATER FLOW RATE (3)		GPM		
AIR COOLED CONDENSER				
UNITS OPERATING				
AMBIENT AIR				
TEMPERATURE		DB/WB		
BAROMETRIC PRESSURE		IN.HG.		
FANS OPERATING		QUAN.		
CHILLED WATER CIRCULATING SYSTEM				
PUMP NO.				
PRESSURE-	IN	PSIG		
	OUT	PSIG		
WATER FLOW RATE (4)		GPM		
WATER FLOW TO USER EQUIPMENT				
TOTAL RATE		GPM		
TEMPERATURE-	TO	DEG. F.		
	FROM	DEG. F.		
WATER FLOW RATE (5)		GPM		
TANK NO.				
FILLING WITH CHILLED WATER				
START, TIME		HR/MIN		
TERMINATE, TIME		HR/MIN		
WATER				
FLOW RATE (4)		GPM		
TEMPERATURE	ENTER	DEG. F.		
	EXITS	DEG. F.		
CHILLED WATER WITHDRAWAL				
START, TIME		HR/MIN		
TERMINATE, TIME		HR/MIN		
FLOW RATE (4)		GPM		
TEMPERATURE	TO	DEG. F.		
	FROM	DEG. F.		

NOTES:
(1) From flow meter or derive from cooler pressure drop/flow curve using pressure differential converted to feet.
(2) From flow meter, derive from condenser pressure drop/flow curve using pressure differential converted to feet or derive from condenser water pump (if one for each condenser) flow/head curve using pump inlet and outlet pressure differential converted to feet.
(3) From flow meter or derive from pump flow/head curve using inlet and outlet pressure differential converted to feet.
(4) From flow meter or derived from brine pump gpm/head curve using pump inlet and outlet differential pressure converted to feet.
(5) Obtain from flow meter or possibly same as pump flow.

The designer should make periodic visits to the site during the initial year of operation to review logs and electric bills and offer additional suggestions to the operating engineer. At least some of these periodic visits should be made during both very hot days and on days with very light loads. Observation of operations during these extremes will provide a valuable insight into the capability and flexibility of the cool storage system.

The involvement of the designer can be reduced considerably after the first year of operation, but will still prove worthwhile. Consideration should be given to having the designer perform a monthly review of the daily logs, recordings, and utility bills, to correlate measured performance with original expectations.

Chapter 10

CASE STUDIES

This section comprises two case studies of actual cool storage installations in commercial buildings. The first relates to an ice storage system; the second to a chilled water storage system.

CASE STUDY 1: ICE STORAGE SYSTEM

A brief analysis conducted in 1979 for a 265,000 ft^2, 24-story San Francisco (CA) office tower indicated that a cool storage system could possibly provide space cooling in the most cost-effective manner (16). Two space cooling systems were designed for purposes of feasibility analysis, one using conventional water chillers, the other using ice storage. (Chilled water cool storage was not considered due to space availability and first-cost disadvantages.) Analysis showed ice storage would be competitive with a conventional system on a first-cost basis, and far more cost-effective on a life-cycle basis due to the "time-of-day" utility rates in effect. The ice storage system is now in place and operating smoothly.

System Selection and Analysis

In conducting the feasibility analysis, a computer program was used to calculate peak and hour-by-hour cooling system design requirements. As shown in Figure 10-1, the peak load was calculated as 460 tons, with storage requirements being calculated as 3300 ton hours.

It was determined that an optimum cool storage plant would use minimum storage and compressor capacity, thus requiring 24-hour refrigerating system operation to meet design load. Accordingly, the storage plant refrigeration system's capacity was determined to be (3,300 ton hours ÷ 24 hours =) 137.5 tons.

In determining storage capacity, it was assumed that the refrigeration plant would provide cooling used directly in the space for ten hours. Accordingly, storage capacity was determined as (14 ÷ 24 x 3,300 ton hours =) 1,925 ton hours. (It was noted that some of the storage requirement could be provided by the system itself, in that coils, pumps, heat exchangers, pipes, valves, fittings, and the water in

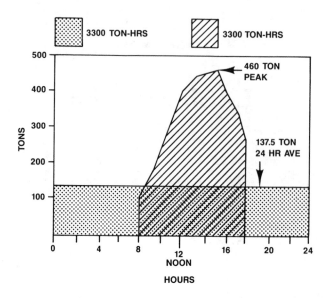

Figure 10-1. Tonnage versus hour-by-hour design load

the system can store thermal energy, in this case, approximately 100 ton hours. However, this was not considered in sizing equipment.)

A comparison of the two systems showed:

	System with Ice Storage	System with Conventional Plant
Compressor Capacity	137.5 tons (483.3 kw)	460 tons (1617 kw)
Cooling Tower Capacity	137.5 tons (483.3 kw)	460 tons (1617 kw)
Chilled Water Capacity	460 tons (1617 kw)	460 tons (1617 kw)

Central Plant. Because some redundancy was desired in the plant, the nominal 137-ton load was split between two 70-ton and fully independent refrigerating and storage circuits, each connected to a 960 ton hour storage system. This approach also optimized the system's ability to minimize electrical demand. Water-cooled heat rejection equipment was selected because the central plant had to be located in the basement, while the cooling towers had to be on the roof.

Air-Handling System. Single-duct, variable-air-volume (VAV) boxes and air-handling units (AHUs) were selected for location on each of the occupied office floors. This was done, in part, to improve life safety aspects by avoiding the use of vertical shafts which are potential paths for smoke migration. Typical floor AHUs used with ice storage had four-row cooling coils and 7-1/2 hp motors, as opposed to the six-row cooling coils and 10 hp motors needed for the conventional system.

Refrigerant System. Given the size of the equipment and the availability of compressors, it was determined that R-22 was the most suitable refrigerant for this system.

Refrigerant Circuit. It was assumed initially that a pumped overfeed-nonexpansion valve system would provide the best refrigerating performance. However, after studying the advantages of subcooling, a suction accumulator/heat exchanger refrigerant system was designed. In this system, liquid leaving the condenser is routed through a heat exchanger which is surrounded by suction gases on their way to the compressor, thus making refrigerant circulating pumps unnecessary.

The ice builder used thermoelectric expansion valves which respond to the presence of liquid refrigerant on their thermister controllers. Their use with a suction accumulator/heat exchanger ("super pot") enhances evaporator performance by allowing the cooled refrigerant a further entry into the 1-1/4-inch steel evaporator tubes before evidence of flashing occurs. A greater percentage of the evaporator surface thus is wetted. This reduces evaporator pressure drop and increases effectiveness of the evaporator area. In addition, subcooling causes more heat transfer because a cool refrigerant is being provided to the evaporator.

Equipment Selection

Factors associated with selection of equipment are as follows.

Refrigerating Compressors. Reciprocating, open, industrial-quality compressors were selected, using standard, open, drip-proof motors with across-the-line starting gear. Design conditions (0.5% wet-bulb, 63F) allowed a maximum condensing temperature of 90F on the average suction temperature upon which ice builder selection was based. At these operating conditions, the kW/ton published by the compressor manufacturer was 0.712 kW/ton, judged adequate from an energy consumption point of view when compared against centrifugal refrigeration.

<u>Cooling Tower and Condensing Pumps</u>. Tower and pump capacity were sized to meet the nominal 140-ton load. This equipment was 30% of the size of the conventional system's towers and pumps.

<u>Water-Cooled Condensers</u>. Condensers were selected to permit design load rejection at a 90F saturated discharge temperature, resulting in selection of a nominal 130-ton condenser for each of the 70-ton loads. Increased condenser surface permitted relatively low head pressure and horsepower requirements. (Since head pressure was not critical to the refrigeration circuit, no operating problems were found.) Close control of superheat was not required since overfeeding refrigerant improves evaporator performance.

Because it was impossible to match equipment exactly to the load, some slight oversizing occurred. However, increased sizing affords protection in the event of unpredictable, larger-than-design loads.

<u>Heat Exchangers</u>. Plate-type (rather than shell-and-tube) heat exchangers were selected because of their reduced space requirements.

<u>Ice Builder</u>. The packaged ice builders selected consist of 1-1/4 inch steel pipe evaporator surface immersed in insulated open tanks. Ice is formed on the outside of the 1-1/4 inch pipe and on the secondary surface that forms baffles to circuit the water as it flows through the tank. The ice builder's controlled water path eliminated the need for agitation to build and melt the ice, thus creating considerable savings (no energy consumption for agitation or for removing heat from the ice builder when ice is being made, and less maintenance requirements).

Each ice builder can store at least 960 ton hours by building approximately two inches of ice on the evaporator's primary and secondary surface, and considering the energy stored in the subcooled ice and water in the ice builder.

<u>Pumping Systems</u>. Water is circulated from the open ice builder tanks through the heat exchanger and is returned to the ice builder, with the closed loop serving the cooling coils providing variable flow. The heat exchangers are piped in parallel on the building side and the flow is varied by cycling of the two pumps to supply the cooling load.

<u>Air-Handling Units (AHUs)</u>. In selecting the AHUs, it was decided to take maximum advantage of the 38F water that could be supplied to the system. Accordingly, face

area was selected to reduce coil face velocity to 350-388 feet per minute. The
combination of increased face area and fewer rows on the coil had a favorable
impact on fan motor size and the static pressure for which the AHUs had to be
selected. Air pressure drop across the cooling coils was 0.18 inches H_2O in the
cool storage system, versus 0.82 inches H_2O in the conventional chilled water
system.

Other Impacts. Structural, electrical and architectural factors also were
affected, as follows:

Structure. When fully charged with water, the ice builders weigh about
172,000 lbs. Accordingly, floor loading in the basement had to be increased
to approximately 580 lb/ft^2. However, the smaller cooling tower permitted use
of less structural steel at the roof level.

Electrical. The cool storage system reduced building electrical demand by
approximately 60%. Further demand reductions were gained due to use of
smaller cooling tower fans, smaller condenser water pumps, and smaller
compressors. Also, improved efficiency is experienced because ice building
occurs at night, when lower wet-bulb temperatures are experienced. It was
projected that the overall utility savings would amount to $38,000 per year
under the present rate structure. Also, reliance on smaller compressors (at
least 160 tons less than those which would have been needed for the
conventional refrigeration plant), and considering the lower cost of service
to fans, towers, pumps, and compressors, resulted in a $60,000 first-cost
savings for electrical requirements.

Architecture. Because the use of ice storage resulted in smaller static
pressure in the fans, less noise would result. This permitted a reduction in
acoustical considerations initially designed into the project to control the
noise of a conventional system.

System Costs

The complete cool storage HVAC and plumbing system for this project cost $2.4
million, $22,000 less than the cost of a system using conventional centrifugal
chillers.

CASE STUDY 2:   CHILLED WATER STORAGE

The office building is located in downtown Baltimore, Maryland. It contains an air conditioned area of approximately 230,000 square feet.

The building is rectangular in shape, with approximately 70,000 square feet of air conditioned area on each of three floors and 20,000 square feet on the fourth floor. Nearly one-third of the air conditioned area is below grade. The perimeter walls of the above-grade portion of the building are constructed with concrete, brick, and glass. The glazed area comprises approximately 25 percent of the exposed wall. All glass is double glazed, solar bronze, except where computer rooms are located on the building perimeter, in which case triple glazing is used to permit a high level of relative humidity to be maintained.

Approximately two-thirds of the building is occupied nine hours per day, five days per week, with the remainder being occupied 24 hours per day, seven days per week with reduced activity on weekends. The portion of the building which is continuously occupied is equipped with computers and other high heat release equipment.

Systems Analysis

Seven basic systems in ten combinations using the TRACE computerized system analysis program were evaluated.

The systems studied included:

- Self-Contained Computer Room Units with Individual Compressors

- Self-Contained Computer Room Units with Chilled Water Coils to be used with a Central Water Chiller

- Variable Air Volume Cooling Units

- Variable Air Volume Heating-Cooling Units

- Incremental Hydronic Heat Pump Units

- Four Pipe Fan Coil Units

- Single Zone Variable Temperature Constant Volume Units

Two chilled water storage systems were evaluated:

- Generation of 100% of required chilled water during off-peak hours

- Continuous generation of chilled water with simultaneous use of stored chilled water during peak hours

Design Criteria. The following design criteria were used in the systems analysis:

a. Summer Conditions

    Outside Air Dry Bulb Temperature: 91F
    Outside Air Wet Bulb Temperature: 77F
    Inside Space Dry Bulb Temperature: 78F
    Inside Maximum Relative Humidity   50%

b. Winter Conditions

    Outside Air Dry Bulb Temperature: 13F
    Inside Space Dry Bulb Temperature: 72F
    Inside Minimum Relative Humidity: 0%
    Minimum Relative Humidity in
        Computer Rooms: 50%

c. Ventilation

    Minimum Outside Air Per Person: 5 CFM
    Toilet Rooms: 2 CFM Per Sq.Ft.

d. Hours of Occupancy

    In general, 9 hours per day, five days per week, with
    one-third of the building being occupied 24 hours per
    day, seven days per week.

e. Building Envelope

    Walls "U" Factor =   0.15
    Roof "U" Factor =   0.08
    Glass, General-Solar Bronze Double "U"   0.60
    Glass, Computer Rooms-Triple Glass "U"   0.36

    Alternate "U" factors for wall, roof, and glass were
    considered as follows:

    Wall and roof "U" factors were set at 0.075 and 0.04
    respectively. The overall effect on the cooling loads
    was a reduction in peak cooling load of approximately
    5% and a reduction in yearly energy consumption for
    cooling of less than 1% due to the large internal loads
    which would be unaffected by a change in building
    envelope. Heating loads would not be significantly
    affected when considered with the system recommended.

    Single glass windows with a "U" factor of 1.1 were
    considered but rejected due to their incompatibility
    with the system of energy recovery recommended. The
    heating load would be increased by approximately 40%,
    and electric draft-barrier heat would be required at a
    much higher outside air temperature.

f. Cooling and Heating Load

    Peak Cooling Load   500 tons
    Peak Heating Load   1,750,000 Btuh

Economic Criteria. The following criteria were used for the economics analysis:

a.   Electric Energy and Demand Costs

Alternate Nos. 1-9: Schedule T, Baltimore Gas and
Electric Company.

Alternate No. 10: Schedule G, Baltimore Gas and
Electric Company

Fuel Adjustment Rate: $0.01067/kWhr (applies to
Schedule T and Schedule G).

Annual Escalation Rate: 4%.

b.   No. 2 Fuel Oil

First Year Cost: $.50/Gallon
Annual Escalation Rate: 4%.

c.   City Water

First Year Cost: $.253/1,000 Gallons
Annual Escalation Rate: 4%.

d.   System Installation and Maintenance Cost

System Type	Installation Cost	Annual Maintenance Cost
VTCV - (Variable Temp., Constant Volume)	$2,700/Ton	$48/Ton
DDVAV - (Double Duct, Variable Air Volume)	$2,700/Ton	$58/Ton
VAV - (Variable Air Volume)	$2,600/Ton	$54/Ton
INCHP - (Incremental Heat Pump)	$2,800/Ton	$48/Ton
S.C.C.R.U. - (Self-Contained Computer Room Unit)	$3,300/Ton [1]	$66/Ton
F.C. - (Fan Coil Units)	$3,500/Ton	$72/Ton
VCTV w/HP - (Variable Temp., Constant Volume Utilized with Heat Pump	$2,800/Ton	$53/Ton
VAVRH - (Variable Air Volume with Reheat)	$2,700/Ton	$56/Ton

[1] ($400/Ton for Electric Included)

e. Additional Costs for Comparison (Not Included in $/Ton)

Alternative	Additional Cost	Additional Costs Include
1	$190,000	Shaft Space, Basement Equipment Space, Tunnel Space
2	$157,500	Shaft Space, Basement Equipment Space, Tunnel Space, Usable Space in Finished Area (Note: less equipment and shaft space required than Alternative #1)
3	$142,500	Same as Alternative #2 with reduced equipment space reflecting deleted water chiller.
4	$142,500	Same as Alternative #3.
5	$167,500	Same as Alternative #1 except less equipment space.
6	$212,000	Same as Alternative #2, but includes additional cost of heating storage and reduced cost of boiler installation.
7	$237,000	Same as Alternative #3, but includes additional cost of cooling storage and reduced cost of water chillers.
8	$ 45,000	Additional space in finished area, and equipment space for chiller.
9	$ 30,000	Additional space in finished area.

10	$237,000	Same as Alternative #7. Although Alternative #7 includes Owner furnished transformers, the electric installation costs were considered equal since the additional cost of transformer vault and feeders under Alternative #10 will offset the additional cost of Owner furnished transformers.

f.   Annual Escalation Rate for Equipment Replacement: 4%.

g.   Mortgage Life: 50 years.

h.   Mortgage Interest: 8%.

i.   Percent Financed: 100%.

j.   Property Tax: 3%.

k.   Income Tax Rate: 0%.

l.   Cost of Capital: 8%.

m.   Depreciation Method, Tax and Book: Straight Line.

Recommended System. The systems analysis indicated that the system using thermal storage for heating and cooling had the lowest annual owning and operating costs of the ten alternatives examined as indicated in Table 10-1.

The costs shown in Table 10-1 show relative cost differences between systems and do not necessarily show complete cost of systems. System No. 1 was disregarded since it did not meet the requirement for individual computer room control.

Table 10-1

Annual Owning and Operating Costs of Alternative Systems

NO.	DESCRIPTION OF ALTERNATIVES	INSTALLED COST	UTILITY COSTS		MAINT. COST YEAR 1	ANNUAL O & O COST
			YEAR 1	YEAR 50		
1	VARIABLE AIR VOLUME HEATING AND COOLING THROUGHOUT	1,272,247	173,114	1,182,937	22,146	864,249
2	VARIABLE AIR VOLUME HEATING AND COOLING PLUS COMPUTER ROOM UNITS WITH CHILLED WATER COILS	1,239,747	177,132	1,210,398	22,146	869,671
3	VARIABLE AIR VOLUME HEATING AND COOLING PLUS COMPUTER ROOM UNITS WITH INDIVIDUAL COMPRESSORS	1,301,694	185,191	1,265,461	24,005	928,872
4	SAME AS ALTERNATIVE #3 EXCEPT USING VARIABLE AIR VOLUME REHEAT FOR HEATING	1,301,694	185,191	1,265,461	23,745	928,078
5	VARIABLE AIR VOLUME INTERIOR SPACES WITH FOUR PIPE FAN COIL UNITS FOR PERIMETER SPACES	1,276,853	185,161	1,265,256	22,149	904,006
6	SAME AS ALTERNATIVE #2 PLUS THERMAL STORAGE FOR HEATING	1,294,247	174,117	1,189,792	22,146	871,947
7	SAME AS ALTERNATIVE #2 PLUS THERMAL STORAGE FOR HEATING AND COOLING	1,319,247	165,021	1,127,637	18,082	837,033
8	INCREMENTAL HYDRONIC HEAT PUMPS PLUS COMPUTER ROOM UNITS WITH CHILLED WATER COILS	1,217,360	201,360	1,375,955	20,728	966,016
9	INCREMENTAL HYDRONIC HEAT PUMPS PLUS COMPUTER ROOM UNITS WITH INDIVIDUAL COMPRESSORS	1,269,630	198,306	1,355,083	22,419	985,165
10	SAME AS ALTERNATIVE #7 WITH BUILDING PURCHASING ELECTRIC POWER ON SCHEDULE G IN LIEU OF SCHEDULE T	1,319,247	183,361	1,252,961	18,082	893,032

System Description

Interior areas of the building are supplied by single duct variable air volume
(VAV) terminal units; perimeter areas are supplied by double-duct VAV units, and
computer rooms are supplied by constant volume, variable temperature packaged
computer room units.

The computer room units are supplied with chilled water at 42F; the single-duct VAV
units and one deck of the double duct units are supplied with air at a constant

55F, and the other deck of the double-duct units is supplied with variable temperature air.

The building's space heating requirements are met by a system that uses heat-reclaiming chillers in conjunction with thermal storage, instead of hot water boilers.

Additionally, thermal storage is used to reduce water chiller capacity and to generate chilled water during off-peak hours.

An electric water heater is used to heat the building to 50F during unoccupied periods. Should the storage tank be out of service, the electric heater can heat the building to 65F during occupied periods, and it can also supplement the heat-recovery chiller when necessary.

Separate air-handling equipment supplies the dual temperature deck and the cold decks of the VAV system. The dual temperature deck is supplied with hot air at outdoor air temperatures below 65F, and with variable temperature cold air at outside air temperatures 65F and above.

The dual temperature deck is supplied by one 45,000 cfm air-handling unit, and the cold deck is supplied by two 45,000 cfm units. Outside air for ventilation is supplied through the cold deck, whereas the dual temperature deck uses 100% recirculated air. Outside air for atmospheric cooling is used by the air handling units supplying the cold deck at outside air temperatures below 55F.

The storage system includes three 40,000 gallon horizontally mounted metal storage tanks. Each tank is divided into four compartments by a system of vertical baffles. The multiple compartments are intended to aid in stratification within the tank and the baffles are intended to reduce stagnation or "dead space" within a compartment, thereby improving efficiency.

As shown in Figure 10-2, cold water enters storage high, and then drops to the bottom of the first compartment, mixing with the water in that tank. The cold water then is forced through the orifice near the bottom of the first baffle, travels upward across the weir of the second baffle into the second compartment, and follows a similar path to additional compartments and additional tanks.

Each compartment contains two thermometer wells -- one high, one low -- drain, air vent, access manhole, and two view ports (one for light) as shown in Figure 10-3.

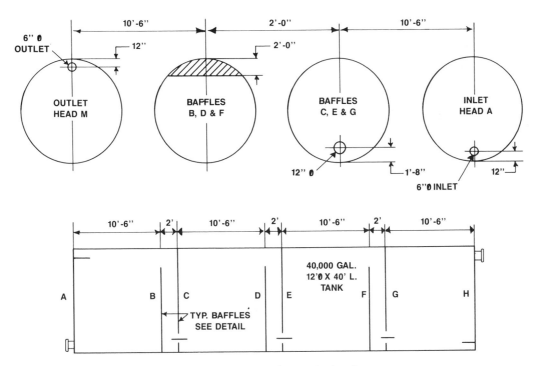

Figure 10-2. Typical storage tank elevation

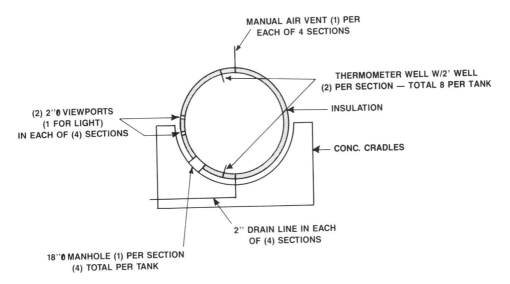

Figure 10-3. Typical cross section of tank

Previous studies using infrared photography and thermistor sensing techniques have indicated that "live" storage capacity as high as 95% of tank capacity can be anticipated through storage tank compartmentization.

The control sequences for the storage system are shown in Figure 10-4 and described below. The control sequence for the thermal storage system with outside air temperature below 65F as indicated on Figure 10-4 is as follows:

a. Secondary chilled water pump supplying variable flow secondary system runs continuously. Primary chilled water pump also runs continuously.

b. Secondary and primary heating water pumps run continuously.

c. Heat recovery chiller is modulated as required to maintain 105F return heating water temperature.

d. Control of the chiller will be overriden by the thermostat in compartment "E" of the storage system whenever the cooling demand exceeds the heating demand, and will run at full capacity until the temperature of chilled water in storage compartment "H" reaches 43F.

e. The cooling tower system will be energized whenever the return heating water temperature reaches 110F, and the electric heater will be energized should the return heating water temperature fall below 100F.

The control sequence for the thermal storage system with outside air temperature 65F or above as shown on Figure 10-4 is as follows:

a. Secondary chilled water pump supplying variable flow secondary system runs continuously.

b. With outside air temperature between 65F and 80F, the chiller (and its associated auxiliary equipment) is started whenever the temperature in storage tank compartment "E" reaches 50F and runs at full capacity until the temperature in compartment "H" reaches 42F. The chiller supplies 41F chilled water.

c. With outside air temperature above 80F, the chiller is cycled off or modulated to match the cooling load when the temperature in compartment "L" reaches 42F, and on at full capacity with temperature in compartment "I" at 50F.

d. The primary chilled water pump provides a constant flow of 350 gpm when running.

e. The secondary chilled water flow to the system varies from approximately 75 gpm minimum flow to 700 gpm during peak cooling demand.

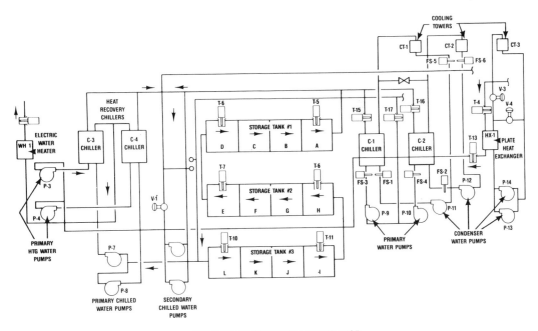

OUTSIDE AIR TEMPERATURE BELOW 65°F
HEATING LOAD EXCEEDS COOLING LOAD
STORAGE TANK CHARGING

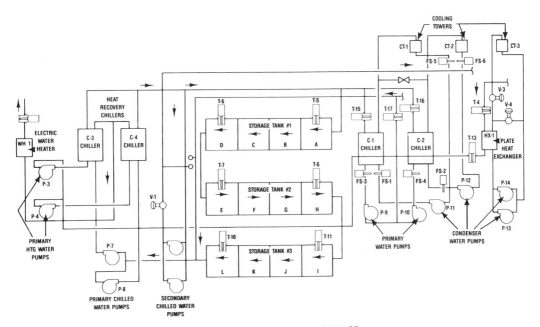

OUTSIDE AIR TEMPERATURE BELOW 65°F
HEATING LOAD EXCEEDS COOLING LOAD
STORAGE TANK CHARGING

Figure 10-4.   Control sequence for storage system

f. Whenever the primary chilled water pump is off, 42F chilled water is supplied from compartment "A" and returned to compartment "L", and system is in full discharge.

g. Whenever the secondary flow matches the primary flow, the storage system is inactive.

h. With the secondary system flowing more water than the primary system, the storage tanks are in a partial discharge condition, and are in a partial charging condition whenever secondary flow exceeds primary flow.

Figure 10-5 illustrates the predicted performance of the storage system for a typical winter and a typical summer week.

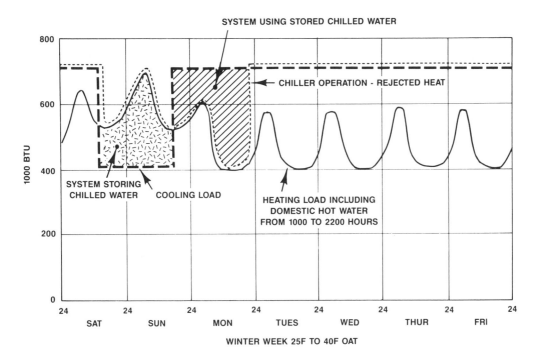

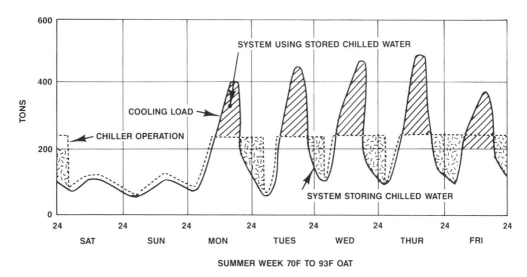

Figure 10-5. Predicted performance of the storage system for a typical winter and summer week

Chapter 11

SOURCES OF INFORMATION
ON COOL STORAGE

Following is a list of selected sources from whom further information about cool storage may be obtained. Utility marketing representatives are likely to be familiar with many additional sources. They should be contacted as well:

ASSOCIATIONS, SOCIETIES AND INSTITUTES

Air-Conditioning and Refrigeration Institute
1815 N. Fort Myer Drive
Arlington, VA 22209
(703) 524-8800. Telex 89-2351

American Consulting Engineers Council
1155 Fifteenth Street, NW
Suite 713
Washington, DC 20005
(202) 296-5390

American National Standards Institute (ANSI)
1430 Broadway
New York, NY 10018
(212) 354-3300

American Public Power Association
2301 M Street, NW
Washington, DC
(202) 775-8300

American Society of Mechanical Engineers
United Engineering Center
345 E. 47th Street
New York, NY 10017
(212) 705-7722

American Society for Testing and Materials
1916 Race Street
Philadelphia, PA 19103
(215) 299-5400

American Society of Heating, Refrigeration and
 Air-Conditioning Engineers, Inc.
1791 Tullie Circle, NE
Atlanta, GA 30329
(404) 636-8400

Association of Energy Engineers
4025 Pleasantdale Road
Suite 340
Atlanta, GA   30340
(404) 447-5083

Edison Electric Institute
1111 19th Street, NW
Washington, DC   20036
(202) 828-7400

Electric Power Research Institute
3412 Hillview Avenue
P.O. Box 10412
Palo Alto, CA   94303
(415) 855-2401

International Institute of Ammonia Refrigeration
111 E. Wacker Drive
Chicago, IL   60601
(312) 644-6610

Mechanical Contractors Association of America, Inc.
5530 Wisconsin Avenue, NW
Suite 750
Washington, DC   20015
(202) 654-7960

National Commercial Refrigeration Sales Association
1900 Arch Street
Philadelphia, PA   19103
(215) 564-3484

Northamerican Heating & Airconditioning Wholesalers Association
1661 W. Henderson Road
Columbus, OH   43220
(614) 459-2100

Refrigeration Engineering Advisory Council
Rural Route 1-Box 123
West Brooklyn, IL   61378

Solar Energy Research Institute
1617 Cole Blvd.
Golden, CO   80401

MANUFACTURERS - ICE BUILDING EQUIPMENT

ACME Div. Heat Transfer Group G&W
3151 W. Michigan Avenue
P.O. Box 88
Jackson, MI
(517) 787-1000

Baltimore Aircoil Co., Inc.
P.O. Box 7322
Baltimore, MD 21227
(301) 799-1300

Calmac Mfg. Corp.
150 S. Van Brunt Street
P.O. Box 710
Englewood, NJ 07631

Caloskills Div.
Girton Mfg. Co.
Millville, PA 17846
(717) 458-5523

Chester-Jensen Co., Inc.
P.O. Box 908
Chester, PA 19016
(215) 876-6276

Flakice Corp.
60 Liberty Street
Metuchen, NJ 08840
(201) 494-1070

Follet Corp.
P.O. Box Drawer D
Easton, PA 18042
(215) 252-7301

Gulf & Western Heat Transfer Group
1625 E Voorhees Street
Danville, IL 61832
(217) 446-3710

Sunwell Engineering Co.
18 Killaloe Rd., No. 4
Concord, Ontario
Canada L4K 1CB
(416) 738-1222

MANUFACTURERS - CONTROLS, ICE BANK

Flakice Corp.
60 Liberty Street
Metuchen, NJ 08840
(201) 494-1070

Hoffman Controls Corp.
2463 Merrell Rd.
Dallas, TX 75229
(214) 243-7424

Honeywell, Inc.
Commercial Construction Div.
Honeywell Plaza
Minneapolis, MN 55408
(612) 870-5200

Motors & Armatures, Inc.
250 Rabro Drive
E. Hauppauge, NY 11788

Ranco Controls
8115 U.S. Route 42N
Plain City, OH 43064

White Rodgers Div.
Emerson Electric Co.
9797 Reavis Road
St. Louis, MO 63123
(314) 577-1300

MANUFACTURERS - STORAGE TANKS

Ace Buehler, Inc.
321 W. Katella Avenue
Orange, CA 92668
(714) 538-8514

Aero Tec Laboratories, Inc.
45 Spear Road
Ramsey, NJ 07446
(201) 825-1400

Bethlehem Steel Corp.
Martin Tower
Bethlehem, PA 18042

Calmac Manufacturing Co.
150 S. Van Brunt Street
P.O. Box 710
Englewood, NJ 07631

Dean Products, Inc.
985 Dean Street
Brooklyn, NY 11238
(212) 789-4444

Federal Boiler Co., Inc.
277 Fairfield Road
Fairfield, NJ 07006
(201) 227-9075

Fiber-Rite Products, Inc.
P.O. Box 38095
Cleveland, OH 44138
(216) 235-6800

Harris Thermal Transfer Products, Inc.
19830 S.W. 102nd Street
Tuatalin, OR 97062

Heil Process Equipment
Xerox Fiberglass Inc.
34250 Mills Road
Avon, OH 44011

National Integrated Systems, Inc.
1656 Sierra Madre Circle
Placentia, CA 92670

Owens-Corning Fiberglass Corporation
Non-Corrosive Products Div.
Fiberglass Tower
Toledo, OH 43659

Paul Mueller Co.
P.O. Box 828
Springfield, MO 65801
(417) 831-3000

Riley-Beaird Div.
United States Riley Corp.
P.O. Box 3115
Shreveport, LA 71130
(318) 865-6351

Smith Precast Inc.
2410 West Broadway
Phoenix, AZ 85041

Thermal Energy Storage, Inc.
10637 Roselle Street
San Diego, CA 92121
(714) 453-1395

LOCAL SOURCES

Societies, Associations and Institutes

Utilities

Building Code Authorities

Architects

Consulting Engineers

Refrigeration Contractors

Suppliers

Dealers

U.S. GOVERNMENT SOURCES

U.S. Department of Energy
1000 Independence Avenue
Washington, DC 20585
(202) 252-1508

Argonne National Laboratory
9700 S. Cass Avenue
Argonne, IL   60439
(312) 972-7108

Oak Ridge National Laboratory
P.O. Box Y
Oak Ridge, TN   37830
(615) 574-0330

Chapter 12

REFERENCES

1. "Medical Arts" Storage-Type Installation Saves 280 Horsepower on Air-Conditioning Equipment Requirements", Your Electric Ally, published by Dallas Power & Light Company, Dallas, Texas, February, 1938

2. A Guide to Ice Storage System Design, Baltimore Air Coil Co. Baltimore, Maryland, 1982.

3. An Introduction to Ice Bank Stored Cooling Systems for Commercial Air Conditioning Applications Calmac Manufacturing Corporation, Englewood, New Jersey

4. Instant Ice from Sunwell, Sunwell Engineering Company, Maple Ontario, Canada

5. Ayres Associates, A Guide for Off-Peak Cooling of Buildings, Southern California Edison Co., Rosemead, California.

6. RCF, Inc. Commercial Cool Storage Primer, Electric Power Research Institute, EPRI EM-3371, January 1984.

7. Enviro-Management & Research, Inc. Characteristics of Commercial Sector Buildings and Their Use and Demand, Electric Power Research Institute, Palo Alto, California (not yet published)

8. ASHRAE Handbook, Fundamentals Volume, 1981 American Society of Heating, Refrigerating and Air-Conditioning Engineers, Inc., Atlanta, Georgia, 1981

9. Domke, Lance J. Design of HVAC Systems with Ice-Based Thermal Storage, presented at ASHRAE Annual Meeting, Washington, DC, June, 1983.

10. Thermal Energy Storage Inducement Program for Commercial Cool Storage, San Diego Gas & Electric, San Diego, California, November, 1983.

11. ASHRAE Handbook, Systems Volume, 1984, American Society for Heating, Refrigerating and Air Conditioning Engineers, Atlanta, Georgia, 1984

12. Tamblyn, Robert T., "Thermal Storage Applications", Heating/Piping/Air Conditioning, January, 1982

13. Thermal Energy Storage: Cooling Commercial Buildings Using Off-Peak Energy Seminar Proceedings, Electric Power Research Institute, Palo Alto, California, EPRI EM-2244, February, 1982.

14. Wildin, Maurice W. Use of Thermally Stratified Water Tanks to Store Cooling Capacity, presentation at ASME Solar Energy Conference, Las Vegas, Nevada, April, 1984

15. <u>Measuring Productivity in Construction, A Construction Industry Cost Effectiveness Project Report</u>, The Business Roundtable, New York, New York, September, 1982

16. Gilbertson, T. A., and R. S. Jandu. "24 Story Tower Air-Conditioning System Employing Ice Storage - A Case History", <u>ASHRAE Transactions</u>, 1984, Vol. 90. Part 1.

Appendix

COOL STORAGE INVESTMENT ANALYSIS

Owners are not willing to invest in a cool storage system without first being convinced that its added cost could be recovered through electric demand and energy cost savings within an acceptable period of time. It is therefore imperative to perform an accurate economic evaluation of each cooling option. This section demonstrates through example the manner in which key factors are analyzed. The example selected is a large office building for which three different cool storage operating modes and three representative commercial-class rate schedules are considered.

To simplify discussion, certain assumptions have been made. As an example, it is assumed that the amount of electricity consumed by a conventional cooling system equals the amount consumed by any of the three cool storage options. This is a reasonable assumption to make given relative performance losses and gains associated with the equipment involved. In essence, the lower cooling temperature required of the refrigeration equipment in storage systems reduces both the capacity and efficiency of the compressor. However, such losses are largely recovered by performance gains derived from operation of refrigeration equipment in cool storage systems at night. Because outdoor temperatures are typically lower during nighttime hours than during daytime hours, the required work of refrigeration is correspondingly reduced.

DESIGN DAY LOAD AND EQUIPMENT CAPACITY

The building's cooling and noncooling loads on the design day (June 21, 1981) are illustrated in Figure A-1. The cooling component of the load was inferred by subtracting the noncooling load as measured on a weather-neutral day, i.e., a day when the cooling system would not be operating. A conventional cooling system is sized to meet the peak cooling load on the design day. As can be seen in Figure A-1, the load which occurs during midday, corresponds to the output of a 700 kilowatt refrigeration system. The capacities of the storage system and the refrigeration equipment for the three cool storage operating modes being considered -- full storage, load-levelling partial storage and demand-limited partial storage -- are derived as follows.

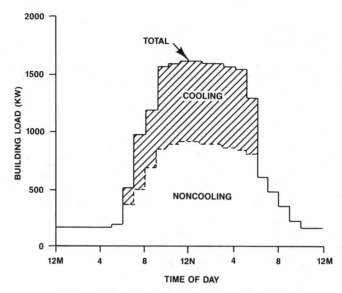

Figure A-1.   Hourly load profile on the design day

Full Storage

Figure 3-2 illustrates the design day load curve under the <u>full storage mode</u> of operation for a case in which the off-peak period extends from 10:00 p.m. to 9:00 a.m. This period corresponds to the off-peak period defined in the time-of-day rate schedule considered below.

The methods used to estimate the required capacities of the refrigeration and storage systems under the full storage mode are similar to those presented under the demand-limited mode of operation, discussed below. In the present example, the capacities are 750 kW for the refrigeration equipment and 6,415 kWh for the storage system.

Load-Levelling Storage

In the load-levelling storage mode, illustrated in Figure A-3, design day cooling load is obtained by taking the hourly difference between the total building load and the building noncooling load and summing this difference over the cooling hours. For the building involved, this works out to be 7,200 kWh which, when divided by 24 hours, yields refrigeration equipment capacity of 300 kW. The building's load profile on the design day for the load-levelling storage mode of operation is obtained by adding the constant compressor load to the noncooling building load.

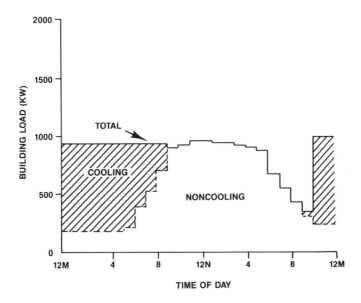

Figure A-2.   Design day load profile under the full storage operating mode

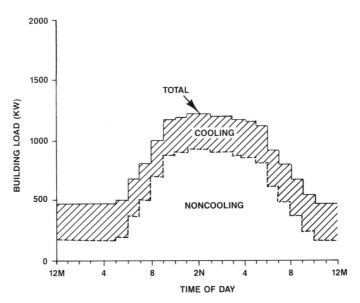

Figure A-3.   Design day load profile under the load-levelling storage operating mode

Because the compressor is rated at 300 kW, the storage requirement is equal to the
cooling load in excess of this capacity summed over all building cooling hours:
3,755 kWh. As indicated in Figures A-1 and A-3, only about 40% of the building
peak-hour cooling load is met directly by the compressor on the design day; the

rest is supplied from storage. The fraction met by the compressor increases on either side of the peak until, during the "shoulder" hours, compressor output exceeds the direct cooling load and part of the compressor output goes into storage. During nighttime hours, the refrigeration equipment is devoted entirely to cooling the storage medium.

Demand-Limited Storage

Figure A-4 illustrates the effect of demand-limited partial storage on the building involved. As can be seen, the building's peak demand has been reduced to the noncooling peak demand, which occurs between the hours of 11:00 a.m. and 1:00 p.m. Sizing a demand-limited partial storage system designed to operate during the shoulder periods requires information about the noncooling load profile as well as about the cooling load profile. full utilization of the shoulder periods will not likely be possible. Use of the "idealized" operating mode will not significantly affect the storage and compressor capacity requirements estimated below because the amount of cooling accomplished during the one or two hours just before and after the peak-load hours is relatively small.

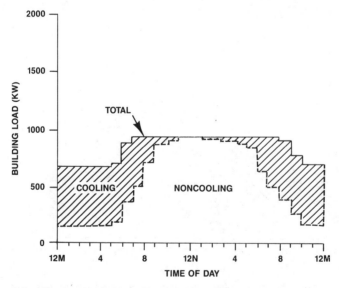

Figure A-4. Design day load profile under the demand-limited partial storage operating mode

The methods used to estimate the required capacities of the refrigeration and the storage systems under the demand-limited mode of operation are straightforward, but more tedious to illustrate than those for the load-levelling mode. Given that the total cooling load on the design day is 7,200 kWh, the problem is to determine the required capacity of the compressor by filling in load on both sides of the non-cooling building load profile until 7,200 kWh have been added. The procedure is illustrated in Figure A-5.

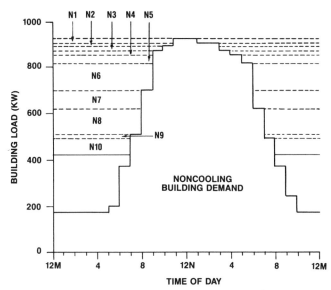

Figure A-5. Graphical procedure for determining compressor capacity in the demand-limited storage operating mode

The highest noncooling demand occurs between 11:00 a.m. and 1:00 p.m. and is 15 kW higher than the next highest noncooling demand which occurs between 1:00 p.m. and 3:00 p.m. Thus, in this example, a cooling load of 15 kW can be applied over 22 hours without creating a new peak. As indicated in Table A-1, the first cooling block N1 contains a "area" of 330 kWh (15 kW x 22 hours). The next block N2 has a height of 10 kW -- the difference between the 1:00 to 3:00 p.m. noncooling demand and the next highest 10:00 to 11:00 a.m. noncooling demand -- and can be applied over 20 hours for a total area of 200 kWh, and a cumulative area for the first two blocks of 530 kWh. The areas of blocks N3 through N9 can be determined through a similar procedure. As indicated in Table A-1, the cumulative area of the first nine blocks is 6,360 kWh and the cumulative height is 425 kW. Because the area of the tenth block, N10, is more than enough to put the cumulative number of kWh above the total cooling load of 7,200 kWh, the height of this block must be adjusted so that the cumulative area just about equals 7,200 kWh. The compressor capacity is given by the cumulative height of all ten blocks, which in this example equals 501 kW. Note that in actual practice blocks of less than 25 kW may not be practical as they will result in inefficient operation of the compressor.

Table A-1

Estimation of the Compressor Size for the Demand-Limited Storage
Mode of Operation

BLOCK	HEIGHT (KW)	WIDTH (HRS)	AREA (KWH)	CUMULATIVE AREA (KWH)	CUMULATIVE HEIGHT (KW)
N1	15	22	330	330	15
N2	10	20	200	530	25
N3	25	19	475	1005	50
N4	15	17	255	1260	65
N5	35	16	560	1820	100
N6	125	15	1875	3695	225
N7	75	14	1050	4745	300
N8	115	13	1495	6240	415
N9	10	12	120	6360	425
N10	76.4	11	840	7200	501.4

To determine the capacity requirement for the storage system, the conventional cooling and demand-limited load profiles can be overlapped graphically as shown in Figure A-6. The storage requirement is equal to the conventional cooling load in excess of the demand-limited cooling load summed over all of the building cooling hours. In the present example, this works out to a storage capacity of 6,100 kWh. Note that the capacity of the storage should in all cases be large enough to meet the building's cooling load during periods when the compressor is not operating.

MONTHLY PEAK DEMANDS

Because electric bills are computed on a monthly basis, it is necessary to develop information about monthly peak demands to compute annual savings to be derived from cool storage. Figure A-7 shows monthly peak demands as recorded for the building involved, as well as month-average building demands, computed by dividing monthly electricity consumption by the number of hours in each month. The curves for monthly peak demands and the month-average demands clearly show the building cooling load build-up during warm-weather months.

To estimate load-levelling storage operating mode reductions in monthly peak demand, assume that the shape of the cooling load profile remains the same on each day of the cooling season as on the design day. Under this assumption, the

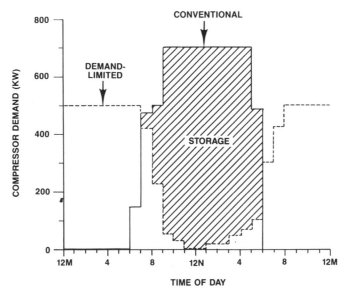

Figure A-6. Overlapping of the demand-limited storage and conventional system load profiles for determining storage capacity in the demand limited storage mode of operation

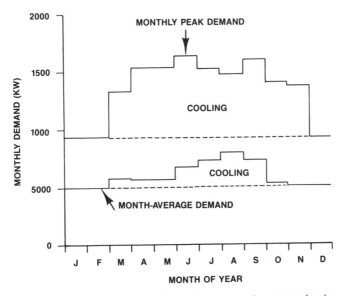

Figure A-7. Monthly peak demands and month average loads

percentage reductions in daily cooling demands achieved with partial storage are the same as on the design day. The design day cooling demand is reduced from 700 kW to 300 kW with partial storage, corresponding to a percentage reduction of 57%. Under the present approach, this same percentage reduction is applied to the peak cooling demand for all other months.

Obviously, for load-levelling partial storage, this approach involves a considerable simplification. Nonetheless, comparisons of bill savings estimated based on this assumption with estimates based on hourly load data over the entire year reveal that this assumption typically produces errors of no more than 15% and, if anything, understates the magnitude of the savings. (The assumption that constant percentage reductions in daily cooling demands leads to an underestimation of bill savings in the case of load-levelling systems can be seen by considering months whose peak-day loads are below the annual peak-day load. Throughout these months it is possible to use the excess storage capacity -- the storage is sized for the annual peak day -- to defer a larger proportion of the cooling load from the building's peak hours to the off-peak hours than possible on the design (annual) peak day. The operating mode would be to enter the start of each daytime period with a fully charged storage tank by running the chiller at rated capacity during the nighttime hours. The chiller would then be operated at a constant, but reduced, level during the daytime hours while the output from storage is carefully prorated to handle the variation in daytime cooling load.)

Since the reduction in design day on-peak cooling demand by both the full and the demand-limited storage systems is essentially 100%, cooling's contribution to monthly peak demands is negligible for all months of the cooling season, assuming proper systems operation. In all cases, the monthly peak demands are established by the building's noncooling loads.

MONTHLY ENERGY CONSUMPTION

The basic assumption underlying the load analyses in this example is that the total quantity of electricity consumed for a conventional cooling system is the same as that used by each of the cool storage systems. For rate schedules that do not incorporate time-of-day energy charges, this assumption represents an adequate specification of building energy use for purposes of monthly bill analysis. If the rate schedule contains a time-of-day energy charge, it will also be necessary to determine the fraction of each month's total energy use that falls within the on- and off-peak periods as defined in the rate schedule.

The first step in dividing the load between peak and off-peak periods is to partition each month's total energy consumption into cooling and noncooling components. In the example analysis, during the month of June, electricity consumption for cooling is 118 mWh and consumption for noncooling uses is 360 mWh.

The next step is to estimate the fraction of each month's load that occurs during peak and off-peak periods. This can be determined from the design day load profile in Figure A-1. For example, if the on-peak period extends from 9:00 a.m. to 10:00 p.m., about 76% of the noncooling load and 85% of the conventional-system cooling load will occur during the on-peak period. If the building operates six days per week, and the sixth day, Saturday, is considered off-peak, these on-peak percentages are reduced to 64% and 70% for the noncooling and cooling components, respectively.

This same method can be used to estimate peak and off-peak cooling loads for the load-levelling, demand-limited, and full storage modes of operation. The percentages of the cooling load occurring during the on-peak period are 0%, 45, and 24, for the full storage, load-levelling and demand-limited storage systems, respectively.

BILL ANALYSIS

Example calculations of the bill savings for cool storage systems are presented for three representative types of commercial-class rate schedules. The distinctive features of the three rate forms are:

- a flat demand charge with ratchet and a declining block energy charge,
- a declining-block energy charge with block widths dependent on billing demand, and
- a seasonally varying demand charge with a time-of-day energy charge.

In addition to these components, most commercial rates also incorporate a customer charge, a fuel adjustment clause and a retail sales tax. The customer charge is usually a flat monthly fee designed to recover fixed equipment and administrative costs associated with providing minimum electric service to a customer. Since the customer charge is the same whether or not a cool storage system is installed, it is not a factor in the estimated bill savings from cool storage.

Neither the fuel adjustment clause nor the retail sales tax is explicitly treated in the example bill analyses, but both are easily incorporated into an actual analysis. The fuel adjustment clause is designed to cover any increases (or decreases) in a utility's direct production-related expenses, and is similar to a constant energy charge per kilowatt-hour. The retail sales tax, which usually is imposed as a constant-percentage tax on the total monthly bill, also can be easily included as a component of the utility bill.

To estimate the annual bill savings for cool storage, it is necessary to calculate the total electric bills for the conventional cooling system and the cool storage system separately and then to take the difference in the bill values. To capture the seasonal variations in building load, the approach adopted here is to calculate monthly bill values over a consecutive 12-month period. The monthly utility bills bills are estimated separately for each cool storage system and for each of the operating strategies under consideration. Note, however, the full storage system was evaluated only for the third rate schedule because it is the only storage schedule incorporating a time-of-day energy charge.

Demand Charge with Ratchet

A feature common to almost all commercial rate schedules is the simple demand charge. Under a demand charge, a customer pays a monthly fee per kilowatt of maximum (or billing) demand. Depending on how high the rate is set per kilowatt, a demand charge can recover part, most or all of the utility investment in plant and equipment. An actual commercial rate schedule incorporating a simple demand charge is reproduced below:

Demand Charge:	$6.58 per kW of Billing Demand
Energy Charges:	First 6,000 kWh @ 4.50¢kWh
	Next 24,000 kWh @ $4.28¢kWh
	More than 30,000 kWh @ 4.02¢kWh

Billing demand is the larger of either (i) the month's peak demand or (ii) 85% of the maximum demand recorded during the previous eleven months during the period June 1 through September 30.

Only commercial-class customers with peak demands above 1000 kW qualify for this particular rate schedule. In the schedule, the demand charge is flat rather than declining-block, and is constant the year round. In other commercial rate schedules of this type, the demand charge may have a declining-block structure and there may be seasonal variations in the values of the demand charges. The example rate also incorporates a declining-block energy charge, but the amounts of energy covered by the first two blocks are so small that almost all of the energy consumed by a large commercial customer will fall into the last block.

The calculation of the electric bills is based on load data derived above. The monthly peak demands and energy consumption levels for the building outfitted with a conventional cooling system are given in Table A-2.

The first step in estimating the demand component of the bill for the conventional cooling system is to calculate billing demand. In computing the monthly values of billing demand, the demand ratchet must be considered. In the example rate schedule, an 85% demand ratchet is in effect, extending over the previous 11 months but applying only to the maximum demand recorded during the period June 1 through September 30. (Other rate schedules of this type may contain several demand ratchets, with one ratchet in effect during the summer months (as in the example rate above), and the other in effect over the rest of the year.

As seen in Table A-2, the ratchet comes into play during the months of November through March, but contributes substantially to billing demand only in December, January and February. The annual demand charge is $115,400. (If the demand ratchet were 100% instead of 85%, the demand component of the annual bill would be $128,403 instead of $115,400, about 11% higher.)

Table A-2

Estimation of Monthly Electric Bills Rate Schedule 1 Mode:
Conventional

BILL COMPONENT	Month												ANNUAL BILL
	JAN	FEB	MAR	APR	MAY	JUN	JUL	AUG	SEPT	OCT	NOV	DEC	
BILLING DEMAND (kW)													
ON-PEAK DEMAND	925	925	1315	1520	1520	1625	1510	1465	1590	1395	1115	925	
RACHET DEMAND	1381	1381	1381	1381	1381	1381	1381	1381	1381	1381	1381	1381	
BILLING DEMAND	1381	1381	1381	1520	1520	1625	1510	1465	1590	1395	1381	1381	
ENERGY CONSUMPTION													
ENERGY (mWh)	372	336	417	400	412	478	537	587	520	388	362	372	
ELECTRICAL BILL ($10^3$$)													
DEMAND CHARGE	9.1	9.1	9.1	10.0	10.0	10.7	9.9	9.6	10.5	9.2	9.1	9.1	115.4
ENERGY CHARGE	15.0	13.6	16.9	16.2	16.7	19.3	21.7	23.7	21.0	15.7	14.6	15.0	209.4
TOTAL	24.1	22.7	26.0	26.2	26.7	30.0	31.6	33.3	31.5	24.9	23.7	24.1	324.8

The effect of load-levelling storage is to reduce the peak cooling loads and, as described above, it reduced monthly peak cooling demand in June from 700 kW to 300 kW, a 57% reduction. By assumption, the peak cooling demands for other months are reduced proportionately. Table A-3 lists the monthly peak demands, energy consumption levels, and bill values for the building incorporating a load-levelling storage system. Notice that the annual electric bill has been reduced from

$324,800 with conventional air conditioning to $298,700 with load-levelling
storage, and that all of this reduction has occurred through reductions in the
demand component of the monthly bills. (It is assumed that, both the conventional
and cool storage systems require the same total amount of electricity each month,
and consequently there are no energy-related bill savings in this example analysis.
The calculation of the energy component of the electric bill could have been
omitted.) Although the reduction in the annual bill is not large -- about 8% --,
it still represents a substantial reduction in that part of the total bill attribu-
table to operation of the air conditioning system.

Table A-3

Estimation of Monthly Electric Bills Rate Schedule 1 Mode:
Load-Levelling Partial Storage

BILL COMPONENT	Month												ANNUAL BILL
	JAN	FEB	MAR	APR	MAY	JUN	JUL	AUG	SEPT	OCT	NOV	DEC	
BILLING DEMAND (kW)													
ON-PEAK DEMAND	925	925	1090	1180	1180	1225	1175	1155	1210	1125	1115	925	
RACHET DEMAND	1041	1041	1041	1041	1041	1041	1041	1041	1041	1041	1041	1041	
BILLING DEMAND	1041	1041	1090	1180	1180	1225	1175	1155	1210	1125	1115	1041	
ENERGY CONSUMPTION													
ENERGY (mWh)	372	336	417	400	412	478	537	587	520	388	362	372	
ELECTRIC BILL ($10^3$$)													
DEMAND CHARGE	6.8	6.8	7.2	7.8	7.8	8.1	7.7	7.6	8.0	7.4	7.3	6.8	89.3
ENERGY CHARGE	15.0	13.6	16.9	16.2	16.7	19.3	21.7	23.7	21.0	15.7	14.6	15.0	209.4
TOTAL	21.8	20.4	24.1	24.0	24.5	27.4	29.4	31.3	29.0	23.1	21.9	21.8	298.7

The monthly bill components for the demand-limited storage system are given in
Table A-4. In this case, electrical requirements for refrigeration do not contribute to monthly peak demands; these demands are established by the building's
noncooling loads. Not surprisingly, because of the seasonal leveling of the peak
demands, the ratchet plays no role in establishing any of the monthly billing
demands. As such, the demand-limited storage system produces the largest savings
-- about 13%.

Average bill savings per kilowatt of demand reduction on the design-day are $65.30
per kilowatt for load-levelling storage ($26,100 ÷ 400 kW) and $60.30 per kilowatt
for demand-limited storage (42,200 ÷ 700 kW). The bill savings per kilowatt are

Table A-4

Estimation of Monthly Electric Bills Rate Schedule 1 Mode:
Demand Limited Storage

BILL COMPONENT	Month												ANNUAL BILL
	JAN	FEB	MAR	APR	MAY	JUN	JUL	AUG	SEPT	OCT	NOV	DEC	
BILLING DEMAND (kW)													
ON-PEAK DEMAND	925	925	925	925	925	925	925	925	925	925	925	925	
RACHET DEMAND	786	786	786	786	786	786	786	786	786	786	786	786	
BILLING DEMAND	925	925	925	925	925	925	925	925	925	925	925	925	
ENERGY CONSUMPTION													
ENERGY (mWh)	372	336	417	400	412	478	537	587	520	388	362	372	
ELECTRIC BILL (10^3\$)													
DEMAND CHARGE	6.1	6.1	6.1	6.1	6.1	6.1	6.1	6.1	6.1	6.1	6.1	6.1	73.2
ENERGY CHARGE	15.0	13.6	16.9	16.2	16.7	19.3	21.7	23.7	21.0	15.7	14.6	15.0	209.4
TOTAL	21.1	19.7	23.0	22.3	22.8	25.4	27.8	29.8	27.1	21.8	20.7	21.1	282.6

slightly greater for load-levelling storage because of the effect of the demand ratchet. With load-levelling storage, the demand ratchet still comes into play during the months of December, January, and February, the 400 kW reduction in peak demand in June producing 360 kW reductions (85% of 400 kW) in billing demand during these winter months. Further seasonal leveling of peak demands eventually leads to a situation where the ratchet no longer affects billing demand, so that further reductions of the summer peak demand have no effect on billing demands during the winter months.

The additional 300 kW reduction in the summer peak demand achieved by demand-limited storage goes beyond the point where the ratchet contributes to billing demand. Thus, billing demands during the months December through February are only 116 kW less than for load-levelling storage. The bill savings during these months per kilowatt of demand reduction on the design day are therefore proportionately less for demand-limited storage than for load-levelling storage.

As observed above, the percentage reductions in total annual electric bills, -- 8% for load-levelling storage and 13% for demand-limited storage -- do not appear to be large. They are put into better perspective when expressed in terms of percentage reductions in the cooling component of the total electric bill. Table A-5 breaks out the contributions of space cooling to total electric bills for the conventional, load-levelling storage, and demand-limited storage systems. The cost of electrical service for space cooling was determined by first subtracting the building cooling loads from the total building loads and calculating the monthly electric bills for these noncooling loads under the same rate schedule used in the foregoing calculations. These new electric bills were then subtracted from the total bills in Tables A-2, A-3, and A-4 to obtain the cost of electric service for the conventional, load-levelling storage, and demand-limited storage systems shown in Table A-5.

The table indicates that the reductions in cooling components of the electric bill are substantial, amounting to 35% for load-levelling storage and 57% for demand-limited storage. Viewed in this context, the cool storage systems represent potentially important cost-saving alternatives to conventional cooling systems.

Table A-5

Cooling and Non-Cooling Contributions to
Total Annual Electric Bill

RATE SCHEDULE 1

BILL COMPONENT	NON-COOLING	CONVENTIONAL		LOAD-LEVELLING		DEMAND-LIMITED	
		COOLING	TOTAL	COOLING	TOTAL	COOLING	TOTAL
DEMAND (10^3\$)	73.2	42.2	115.4	16.1	89.3	0.0	73.2
ENERGY (10^3\$)	177.2	32.2	209.4	32.2	209.4	32.2	209.4
TOTAL (10^3\$)	250.4	74.4	324.8	48.3	298.7	32.2	382.6
PERCENT REDUCTION IN:							
COOLING BILL				35		57	
TOTAL BILL					8		13

Demand-Dependent Declining-Block Energy Charge

Instead of a simple declining-block energy charge with fixed block widths, many commercial rate schedules contain declining-block energy charges with the block widths dependent on the customer's billing demand. Under this type of schedule, the amount of energy within each block is directly proportional to the customer's billing demand and therefore can vary from month to month. The essential elements of an actual commercial rate of this type are reproduced below:

```
Demand Charge:              $.35 per kW of Billing Demand

Energy Charge:;
  (July through Sept.)      First 250 kWh's per kW @ 6.48¢/kWh
                            More than 250 kWh's per kW @ 2.98¢/kWh

  (All Other Months)        First 175 kWh's per kW @ 6.48¢/kWh
                            More than 175 kWh's per kW @ 2.89¢/kWh
```

Billing demand is larger of either (i) the month's peak demand or (ii) 100% of the maximum demand recorded during the previous 11 months.

Both the summer and winter (nonsummer) energy charges in this rate schedule exhibit the demand-dependent declining-block structure. During the summer, only that monthly consumption in excess of 250 kWh per kW of billing demand qualifies for the tail-block rate; during the winter, the breakpoint falls to 175 kWh per kW of billing demand. In addition to the energy charge, the rate also contains a small demand charge applicable year-round.

The high initial-block rate in this type of schedule is designed to recover not only the direct production-related expenses (fuel and variable operations and maintenance costs) but also part of the utility investment in plant and equipment required to meet the customer's peak demand. The lower tail-block rate reflects the lower average production costs associated with meeting customer loads that are relatively constant over time as opposed to loads that are relatively peaked. (This can be seen more directly by reinterpreting the kWh to kW ratios in the energy charge schedules in terms of monthly load factors -- that is, in terms of the ratio of the month-average rate of electricity consumption to the peak demand. For the 250 kWh per kW ratio specified in the summer schedule, the implied load factor over a 30-day month is .35 [250 kWh/(30 x 24 hr)]; for the 175 kWh per kW ratio in the winter schedule the implied load factor is .24.)

For customers facing this type of declining-block energy charge, the potential bill savings from reducing billing demand can be significant. As an illustration of this, consider a building in which the ratio of monthly consumption to billing demand during a summer month exceeds 250 kWh per kW, putting part of the building's energy consumption into the tail-block. Then the net effect of reducing billing demand by 1 kW, while keeping monthly consumption constant, is to shift 250 kWh from the initial to the tail block for a monthly bill savings of $8.98. The corresponding bill saving per kW reduction under the winter schedule is $6.28.

The change in the monthly bill due to a 1 kW change in billing demand may be interpreted as an "effective" monthly demand charge. The effective summer demand charge of $9.33 per kW under this example rate schedule is significantly greater than the flat demand charge of $6.58 per kW; the effective demand charge under the winter schedule is approximately equal to the flat demand charge. (The summer-month saving is 250 x (6.48¢ - 2.89¢) = $8.98, plus $0.35 for the fixed demand charge, for a total of 9.33. The saving during each winter month is 175 x (6.48¢ - 2.89¢) = $6.28, plus $0.35, for a total of 6.63. Thus, the annual savings over the three summer and nine winter months are $87.66.

A graphical method of estimating the monthly bill is illustrated in Figure A-8. The figure shows the utility bill (expressed in dollars per kW) as a function of the ratio of kWh's of monthly consumption to kW's of billing demand for both the summer and winter energy-charge schedules. Below 175 kWh per kW, both the summer and winter curves overlap, and the slope of both curves equals the initial-block charge (6.48 ¢/kWh), which is the same in both schedules. Above 175 kWh per kW the slope of the winter curve drops to the tail block charge (2.89¢/kWh); the slope of the summer curve also drops to this value but only beyond 250 kWh per kW.

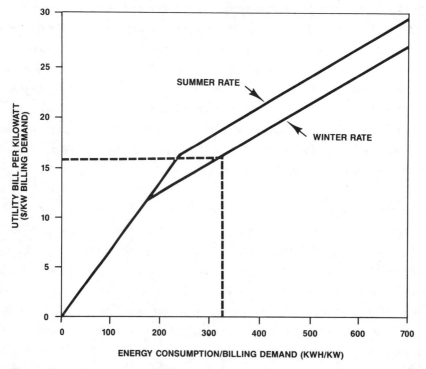

Figure A-8.   Utility bill per kilowatt as a function of the kwh to kw ratio, for the summer and winter energy charges in rate schedule 2

To understand the method of constructing the plot, and how to use it, consider the case of a building with 300 kWh of monthly consumption per kW of billing demand during a winter month. By reading from the winter curve in Figure A-8, the electric bill per kW is found to be $15.82, of which $11.34 is the contribution of the 175 kWh in the initial block of the schedule and $4.48 is the contribution of the 155 kWh in the second block. The total energy component of the monthly bill is then obtained by multiplying the utility bill per kW by the billing demand.

Although the curves in Figure A-8 are specific to the example rate schedule, similar curves can be prepared for any schedule incorporating demand-dependent energy charges. Once the curves characterizing a particular rate schedule have been plotted, the utility bill per kW for any month can be read directly from the plot.

The remainder of this section presents example calculations of the bill savings available for load-levelling and demand-limited cool storage systems. Table A-6 lists the monthly peak-demand and energy-consumption data for the conventional cooling system. Because the example rate schedule contains a 100% demand ratchet over the previous eleven months, monthly billing demand remains constant throughout the year at the annual peak demand of 1,625 kW recorded during June.

Table A-6

Estimation of Monthly Electric Bills Rate Schedule 2 Mode:
Conventional

BILL COMPONENT	Month												ANNUAL BILL
	JAN	FEB	MAR	APR	MAY	JUN	JUL	AUG	SEPT	OCT	NOV	DEC	
BILLING DEMAND (kW)													
PEAK DEMAND	925	925	1315	1520	1520	1625	1510	1465	1590	1395	1115	925	
RACHET DEMAND	1625	1625	1625	1625	1625	1625	1625	1625	1625	1625	1625	1625	
BILLING DEMAND	1625	1625	1625	1625	1625	1625	1625	1625	1625	1625	1625	1625	
ENERGY CONSUMPTION (mWh)													
ENERGY (mWh)	372	336	417	400	412	478	537	587	520	388	362	372	
kWh/kW RATIO	229	207	257	246	254	294	330	361	320	239	223	229	
ELECTRIC BILL (10³$)													
DEMAND CHARGE	.6	.6	.6	.6	.6	.6	.6	.6	.6	.6	.6	.6	6.8
ENERGY CHARGE	21.0	19.9	22.3	21.8	22.1	28.4	30.1	31.5	25.2	21.4	20.7	21.0	285.4
TOTAL	21.6	20.5	22.9	22.4	22.7	29.0	30.7	32.1	25.8	22.0	21.3	21.6	292.2

Table A-6 also lists the monthly ratios of energy consumption to billing demand. The ratios range from a high of 361 kWh per kW during August to a low of 207 kWh per kW during February. A consequence of the 100% demand ratchet is to depress significantly the kWh to kW ratios during winter months compared to the values had there been no applicable demand·ratchet. The ratio in February, for example, would have been 363 kWh per kW instead of 207 kWh per kW if it had been based on monthly peak demand instead of billing demand. As a result of the 100% demand ratchet, almost all of the building's energy consumption during the months November through February falls into the initial block of the energy charge schedule.

Similar monthly breakdowns of peak demand, billing demand, building energy consumption, the kWh to kW ratio, and the demand and energy components of the utility bill are presented in Tables A-7 and A-8 for the cases in which the office building is equipped with load-levelling and demand-limited storage systems, respectively.

Table A-7

Estimation of Monthly Electric Bills Rate Schedule 2 Mode:
Load-Levelling Storage

BILL COMPONENT	Month												ANNUAL BILL
	JAN	FEB	MAR	APR	MAY	JUN	JUL	AUG	SEPT	OCT	NOV	DEC	
BILLING DEMAND (kW)													
PEAK DEMAND	925	925	1090	1180	1180	1225	1175	1155	1210	1125	1115	925	
RACHET DEMAND	1225	1225	1225	1225	1225	1225	1225	1225	1225	1225	1225	1225	
BILLING DEMAND	1225	1225	1225	1225	1225	1225	1225	1225	1225	1225	1225	1225	
ENERGY CONSUMPTION (mWh)													
ENERGY (mWh)	372	336	417	400	412	478	537	587	520	388	362	372	
kWh/kW RATIO	304	274	340	327	336	390	438	479	424	317	296	304	
ELECTRIC BILL ($10^3$$)													
DEMAND CHARGE	.4	.4	.4	.4	.4	.4	.4	.4	.4	.4	.4	.4	5.1
ENERGY CHARGE	18.5	17.4	19.7	19.3	19.6	24.8	26.5	28.0	22.7	18.9	18.2	18.5	252.1
TOTAL	18.9	17.8	20.1	19.7	20.0	25.2	26.9	28.4	23.1	19.3	18.6	18.9	257.2

As described earlier in the Section, the monthly peak cooling demand in June is reduced by 400 kW for partial storage and 700 kW for demand-limited storage. Because the demand ratchet is 100% and extends over eleven months, these peak demand reductions in June translate into billing demand reductions of equal magnitude throughout the year. The percent reductions in total billing demand are 25%

(400 kW ÷ 1625 kW) for load-levelling storage, and 43% (700 kW ÷ 1625 kW) for demand-limited storage.

Table A-8

Estimation of Monthly Electric Bills Rate Schedule 2 Mode:
Demand-Limited Storage

BILL COMPONENT	Month												ANNUAL BILL
	JAN	FEB	MAR	APR	MAY	JUN	JUL	AUG	SEPT	OCT	NOV	DEC	
BILLING DEMAND (kW)													
PEAK DEMAND	925	925	925	925	925	925	925	925	925	925	925	925	
RACHET DEMAND	925	925	925	925	925	925	925	925	925	925	925	925	
BILLING DEMAND	925	925	925	925	925	925	925	925	925	925	925	925	
ENERGY CONSUMPTION (mWh)													
ENERGY (mWh)	372	336	417	400	412	478	537	587	520	388	362	372	
kWh/kW RATIO	402	363	451	432	445	517	581	635	562	419	391	402	
ELECTRIC BILL (10³$)													
DEMAND CHARGE	.3	.3	.3	.3	.3	.3	.3	.3	.3	.3	.3	.3	3.9
ENERGY CHARGE	16.6	15.5	17.9	17.4	17.7	22.1	23.8	25.3	20.8	17.0	16.3	16.6	227.0
TOTAL	16.9	15.8	18.2	17.7	18.0	22.4	24.1	25.6	21.1	17.3	16.6	16.9	230.9

Since monthly energy consumption is assumed to be the same for conventional cooling and both of the cool storage options, any increases in the kWh to kW ratios under partial and full storage come about in this illustration only through reductions in billing demand. As noted above, the percent reduction in billing demand is uniform throughout the year and, as a result, the percent increase in the monthly kWh to kW ratios is also uniform throughout the year. For the 25% reduction in billing demand under load-levelling storage, the corresponding percent increase in the monthly kWh to kW ratios is 33%, and for the 43% reduction under demand-limited storage the percent increase in the ratios is 76%. The net result of increasing the kWh to kW ratios is to shift significant fractions of monthly energy consumption from the initial to the tail block in the energy charge schedule. This, in turn, leads to the lower energy-related monthly bills for load-levelling and demand-limited storage shown in Tables A-7 and A-8. The total annual electric bill is reduced from $292,200 with conventional cooling to $257,200 with load-levelling storage (12% reduction) and to $230,900 with demand-limited storage (21% reduction).

The average bill savings per kilowatt of demand reduction on the design-day are about $89/kW for both the load-levelling and demand-limited systems. These savings are about 50% greater than the savings calculated under the flat demand charge. The higher bill savings stem partly from the effect of the 100% demand ratchet under the present rate, which propagates the billing demand reductions on a one-for-one kW basis throughout the year. Another cause of the higher bill savings is the effective demand charge of $9.33 per kilowatt applicable under the summer schedule, which is considerably higher than the flat demand charge of $6.58 per kW under the previous rate.

Table A-9 breaks out the contributions of space cooling to the total electric bills for the conventional, load-levelling storage, and demand-limited storage systems. The cost of electrical service for space cooling was determined by first subtracting the building cooling loads from the total building loads and calculating the monthly electric bills for the noncooling loads under the same rate schedule used in the foregoing calculations. These new electric bills were then subtracted from the total bills in Tables A-6, A-7, and A-8 to obtain the cost of electric service for the conventional, load-levelling storage, and demand-limited storage systems shown in Table A-9.

Table A-9

Cooling and Non-Cooling Contributions to
Total Annual Electric Bill

BILL COMPONENT	NON-COOLING	CONVENTIONAL		LOAD-LEVELLING		DEMAND-LIMITED	
		COOLING	TOTAL	COOLING	TOTAL	COOLING	TOTAL
DEMAND ($10^3$$)	3.9	2.9	6.8	1.2	5.1	0	3.9
ENERGY ($10^3$$)	203.8	81.6	285.4	48.3	252.1	23.2	227.0
TOTAL ($10^3$$)	207.7	84.5	292.2	49.5	257.2	23.2	230.9
PERCENT REDUCTION IN:							
COOLING BILL				41		73	
TOTAL BILL					12		21

As indicated in Table A-9, almost the entire annual electric bill in each of the three cases is energy-charge related. The demand-charge component of the annual bill is less than 2.5% of the total bill for conventional cooling and is an even

smaller percentage of the annual bill for both of the storage options. The table'
indicates that the reductions in the cooling components of the electric bill are
substantial, amounting to 41% for load-levelling storage and 73% for demand-limited
storage.

Time-of-Day Demand and Energy Charges

In addition to seasonally varying demand and energy charges, many of the newer
commercial rate schedules also contain time-of-day components in either or both the
demand and energy charges. An actual commercial rate schedule of this type is
illustrated below:

    Demand Charge:

        (1)   June through August $9.74 per kW of Monthly
                                          Peak Demand

        (2)   Sept. through May $7.61 per kW of Monthly
                                          Peak Demand

    Energy Charge: (Year-Round)

        On-peak   5.466 ¢ per kWh
        (9:00 am-10:00 pm weekdays)

        Off-peak   2.786 ¢ per kWh
        (10:00 pm-9:00 am weekdays
         plus weekends and holidays)

In this rate, only the energy charge incorporates a time-of-day component. The on-
peak period, extending from 9:00 a.m. to 10:00 p.m. weekdays, covers the period
during which the utility experiences its highest demand. The off-peak period,
extending from 10:00 pm to 9:00 am on weekdays and including the entire day on
weekends and holidays, is a period during which demand is much lower. The energy
charge in effect during the on- or off-peak period is designed to recover the
utility's fuel and variable operation and maintenance costs associated with meeting
electricity use during that period. In the example rate, the on-peak energy charge
is approximately 100% higher than the off-peak rate, reflecting the utility's much
greater reliance on less-efficient, oil-fired, generating plant during the peak
period.

The demand charge is seasonally-dependent but does not contain time-of-day
components. The rate also does not contain a demand ratchet and, as a result, the
demand charge is applied directly to each month's peak electric demand. In other
rate schedules of this type the demand charge may contain time-of-day components --

with the charge per kW normally set high during the on-peak period, and much lower during the off-peak period.

To estimate the energy component of the utility bill under a time-of-day energy charge, monthly energy consumption must be determined separately for the on- and off-peak periods. Knowing the on- and off-peak monthly consumption figures, the energy component of the monthly bill can be calculated by simply multiplying the on- and off-peak consumption levels by the corresponding energy charges. Table A-10 lists the on- and off-peak monthly consumption levels and the energy component of the monthly bill for the commercial office building discussed earlier, for the case in which it is equipped with a conventional cooling system. The monthly bill breakdown in Table A-10 also includes estimates of the demand component of the electric bill. Under this example rate schedule, energy charges account for approximately two-thirds of the total annual bill.

Table A-10

Estimation of Monthly Electric Bills Rate Schedule 3 Mode: Conventional

BILL COMPONENT	Month												ANNUAL BILL
	JAN	FEB	MAR	APR	MAY	JUN	JUL	AUG	SEPT	OCT	NOV	DEC	
BILLING DEMAND (kW)													
PEAK DEMAND	925	925	1315	1520	1520	1625	1510	1465	1590	1395	1115	925	
ENERGY CONSUMPTION (mWh)													
ON-PEAK ENERGY	237	214	269	257	265	312	353	388	342	248	231	237	
OFF-PEAK ENERGY	135	122	148	143	147	166	184	199	178	140	131	135	
ELECTRIC BILL (10^3)													
DEMAND CHARGE	7.0	7.0	10.0	11.6	11.6	15.8	14.7	14.3	12.1	10.6	10.4	7.0	132.1
ENERGY CHARGE	16.7	15.0	18.8	18.0	18.5	21.6	24.3	26.7	23.6	17.4	16.2	16.7	233.5
TOTAL	23.7	22.0	28.8	29.6	30.1	37.4	39.0	41.0	35.7	28.0	26.6	23.7	365.6

Table A-11 presents a monthly breakdown of on- and off-peak energy use and the demand and energy components of the monthly bill for the full storage option.

Similar breakdowns of monthly demand, monthly on- and off-peak energy use, and the demand and energy components of the monthly bill, are provided in Tables A-12 and A-13 for the cases in which the commercial office building is equipped with either a load-levelling or demand-limited cool storage system. The monthly on- and off-peak energy data shown in the tables for load-levelling and full storage were derived using the approximate methods discussed earlier.

Table A-11

Estimation of Monthly Electric Bills Rate Schedule 3 Mode:
Full Storage

BILL COMPONENT	JAN	FEB	MAR	APR	MAY	JUN	JUL	AUG	SEPT	OCT	NOV	DEC	ANNUAL BILL
BILLING DEMAND (kW)													
PEAK DEMAND	925	925	1090	1180	1180	1225	1175	1155	1210	1125	1115	925	
ENERGY CONSUMPTION (mWh)													
ON-PEAK ENERGY	237	214	257	247	254	281	309	330	299	244	230	237	
OFF-PEAK ENERGY	135	122	160	153	158	197	228	257	221	144	132	135	
ELECTRIC BILL (10³)													
DEMAND CHARGE	7.0	7.0	8.3	9.0	9.0	11.9	11.4	11.2	9.2	8.6	8.5	7.0	108.1
ENERGY CHARGE	16.7	15.1	18.5	17.8	18.3	20.8	23.2	25.2	22.5	17.3	16.2	16.7	228.3
TOTAL	23.7	22.1	26.8	26.8	27.3	32.7	34.6	34.6	31.7	25.9	24.7	23.7	336.4

Table A-12

Estimation of Monthly Electric Bills Rate Schedule 3 Mode:
Load-Levelling Storage

BILL COMPONENT	JAN	FEB	MAR	APR	MAY	JUN	JUL	AUG	SEPT	OCT	NOV	DEC	ANNUAL BILL
BILLING DEMAND (kW)													
PEAK DEMAND	925	925	925	925	925	925	925	925	925	925	925	925	
ENERGY CONSUMPTION (mWh)													
ON-PEAK ENERGY	237	214	248	239	246	257	276	288	267	241	230	237	
OFF-PEAK ENERGY	135	122	169	161	166	221	216	299	253	147	132	135	
ELECTRIC BILL (10³)													
DEMAND CHARGE	7.0	7.0	7.0	7.0	7.0	9.0	9.0	9.0	7.0	7.0	7.0	7.0	90.0
ENERGY CHARGE	16.7	15.1	18.3	17.5	18.1	20.2	21.1	24.1	21.6	17.3	16.2	16.7	224.9
TOTAL	23.7	22.1	25.3	24.5	27.1	29.2	30.1	33.1	28.6	24.3	23.2	23.7	314.9

Table A-13

Estimation of Monthly Electric Bills Rate Schedule 3 Mode:
Demand-Limited Storage

BILL COMPONENT	Month												ANNUAL BILL
	JAN	FEB	MAR	APR	MAY	JUN	JUL	AUG	SEPT	OCT	NOV	DEC	
BILLING DEMAND (kW)													
PEAK DEMAND	925	925	925	925	925	925	925	925	925	925	925	925	
ENERGY CONSUMPTION (mWh)													
ON-PEAK ENERGY	237	214	237	229	237	229	237	237	229	237	229	237	
OFF-PEAK ENERGY	135	122	180	171	175	249	300	350	291	151	133	135	
ELECTRIC BILL (10^3)													
DEMAND CHARGE	7.0	7.0	7.0	7.0	7.0	9.0	9.0	9.0	7.0	7.0	7.0	7.0	90.0
ENERGY CHARGE	16.7	15.1	18.0	17.3	17.8	19.5	21.3	22.7	20.6	17.2	16.2	16.7	219.1
TOTAL	23.7	22.1	25.0	24.3	24.8	28.5	30.3	31.7	27.6	24.2	23.2	23.7	309.1

Under conventional cooling, approximately 85% of the design-day cooling load of the office building occurs during the utility's on-peak period. For the three storage options, the percentages of the design day cooling load occurring during the on-peak period are 0% for full storage, 54% for load-levelling storage, and 28% for demand-limited storage. The percentages are then applied to the total monthly energy use from the on- to off-peak periods under full storage load-levelling and demand-limited storage.

Table A-14 presents a summary breakdown of the annual electric bill under the example rate schedule for conventional cooling and for the three cool storage options. As in the preceding bill analyses, the annual bill has been divided into the noncooling and cooling components. To determine the contribution of the noncooling load to the annual bill, a separate bill calculation, based only on the noncooling portion of the building's load, was made. The cooling contribution was taken to be the difference between the annual bills for the total building load and the noncooling load.

As indicated in Table A-14, energy charges account for approximately two-thirds of the annual electric bills for the conventional and the cool storage systems. However, the bill savings are achieved mainly through reductions in demand charges. Reductions in the demand component of the electric bill account for just over 24% of the savings for full storage and just over 80% of the savings for load-levelling and demand-limited storage.

Table A-14

Cooling and Non-Cooling Contributions to
Total Annual Electric Bill

BILL COMPONENT	NON-COOLING	CONVENTIONAL COOLING	CONVENTIONAL TOTAL	FULL COOLING	FULL TOTAL	LOAD-LEVELLING COOLING	LOAD-LEVELLING TOTAL	DEMAND-LIMITED COOLING	DEMAND-LIMITED TOTAL
DEMAND (10^3\$)	90.0	42.1	132.1	0	90.0	18.1	108.1	0	90.0
ENERGY (10^3\$)	196.8	36.7	233.5	22.3	219.1	31.5	228.3	28.1	224.9
TOTAL (10^3\$)	286.8	78.8	365.6	22.3	309.1	49.6	336.4	28.1	314.9
PERCENT REDUCTION IN:									
COOLING BILL				72		37		64	
TOTAL BILL					15		8		14

The percentage reductions of the total annual bill are 15% for full storage, 8% for load-levelling storage, and 14% for demand-limited storage. The percentage reductions in the cooling component of the annual electric bill are considerably larger: 72% for full storage, 37% for load-levelling storage, and 64% for demand-limited storage. The bill savings per kilowatt of demand reduction on the design day are approximately \$81 for full storage and \$73 for load-levelling and demand-limited storage. The bill savings for full storage are slightly greater because of the additional energy-charge savings achieved through the displacement of the entire cooling load to the off-peak period.

EQUIPMENT SIZE AND COSTS

Table A-15 lists the capacities and costs of the conventional air conditioning system and each of the cool storage systems used in the example. The costs are based on the estimates established through methods discussed in Section 3. Also listed in Table A-15 are the net investment costs of each storage system (cost of the storage system less the cost of the conventional cooling system) expressed on a dollars per kilowatt of peak demand reduction basis. Notice that under the load-levelling storage system there are significant cost savings in the chiller component.

Table A-15

Cooling System Size and Cost

RATE/ COOLING MODE	COOLING SYSTEM SIZE				COOLING SYSTEM COSTS (10³$)			KILOWATTS SHIFTED	KWH STORAGE PER KW	NET INVESTMENT COST PER KW SHIFTED ($/KW)		
	CHILLER		STORAGE									
	TONS	KW	10³ GAL	(KWH)	CHILLER	STORAGE	TOTAL	(KW)	SHIFTED	CHILLER	STORAGE	TOTAL
CONVENTIONAL	700	700	-	-	174.9	-	174.9	-	-	-	-	-
FULL	750	750	641.5	6415	184.0	257.8	441.8	700	9.2	13.0	368.3	381.3
LOAD-LEVELLING	300	300	375.5	3755	97.7	162.2	259.9	400	9.4	-193.0	405.5	212.5
DEMAND-LIMITED	500	500	610.0	6100	136.4	246.8	383.3	700	8.7	-55.0	352.6	297.6

ECONOMIC   ANALYSIS

Having estimates of the system investment costs (Table A-15) and the annual
electric bills under the three electric rate schedules (Tables A-5, A-9 and A-14),
the payback periods and net present values of the storage systems can now be
calculated.

Table A-16

Payback on Investment in Cool Storage

RATE/ COOLING MODE	ANNUAL UTILITY BILL CHARGES (10³$)			KILOWATTS SHIFTED (KW)	ANNUAL BILL SAVINGS PER KW SHIFTED ($/KW)			NET INVESTMENT COST/PER KW SHIFTED ($/KW)	PAYBACK (YRS)
	DEMAND	ENERGY	TOTAL		DEMAND	ENERGY	TOTAL		
RATE 1:									
CONVENTIONAL	115.4	209.4	324.8	-	-	-	-	-	-
LOAD-LEVELLING	89.3	209.4	298.7	400	65.3	-	65.3	212.5	3.3
DEMAND-LIMITED	73.2	209.4	282.6	700	60.3	-	60.8	297.6	4.9
RATE 2:									
CONVENTIONAL	6.8	285.4	292.2	-	-	-	-	-	-
LOAD-LEVELLING	5.1	252.1	257.2	400	4.2	83.3	87.5	212.5	2.4
DEMAND-LIMITED	3.9	227.0	230.9	700	4.2	83.3	87.5	297.6	3.4
RATE 3:									
CONVENTIONAL	132.1	233.5	365.6	-	-	-	-	-	-
FULL	90.0	219.1	309.1	700	60.1	20.6	80.7	381.3	4.7
LOAD-LEVELLING	108.1	228.3	336.4	400	60.0	13.0	73.0	212.5	2.9
DEMAND-LIMITED	90.0	224.9	314.9	700	60.1	12.3	72.4	297.6	4.1

Comparison of Payback Periods

Table A-16 shows the payback periods for the full storage, load-levelling, and demand-limited storage modes of operation under the three rate schedules. Note, the full storage system was evaluated only for the third rate schedule.

The rather substantial difference between the payback periods for systems designed for the full storage mode of operation and those designed for the load-levelling or demand-limited modes of operation stems from the larger investment outlay required for the full storage and demand-limited systems. As shown in Table A-16, the storage capacity requirement is greater and the reductions in refrigeration capacity are smaller than for the load-levelling mode of operation. These disadvantages are only partly made up by the greater electric bill savings under the demand-limited mode of operation. Thus, under an investment criterion based on simple payback, the load-levelling storage system emerges as the preferred system.

Net Present Values

Because different investors may assign different discount rates to future bill savings, it is useful to examine how the net present values of alternative investments depend on the discount rate. Figures A-9a, b, and c show the dependence for the different storage systems under the three rate schedules. Here, for purposes of illustration, it has been assumed that the annual electric bill savings, expressed in constant dollars, remain constant over the assumed 20-year lifetime of each investment. This, of course, is equivalent to assuming that the bill savings, measured in ten-year nominal dollars, will increase at the general inflation rate. The discount rate in Figures A-9a, b, and c represents the real cost of money, exclusive of all inflation effects. (Although the cost of money will vary from investor to investor and even from project to project, it should be pointed out that real discount rates above 10% per year are generally regarded as high. For purposes of this analysis an after-tax value of 3.2% and a before-tax value of 6.4% was used. Because electric bill savings represent a reduction in a before-tax expense, a before-tax discount rate should be used in evaluating the cool-storage systems.

The important point illustrated in Figures A-9a, b, and c is that, for this example application, the demand-limited storage system has a higher net present value than the load-levelling storage system for sufficiently low discount rates. In Figure A-9c, the net present value curves for the full and demand-limited systems nearly overlap for low discount rates indicating that full storage is not a clear choice over demand-limited storage even at very low discount rates.

(a) Rate form 1

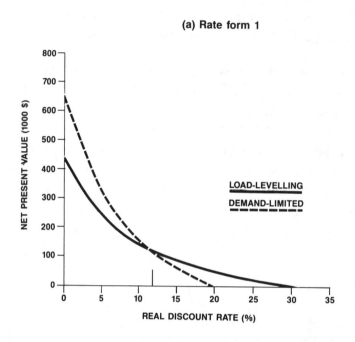

(b) Rate form 2

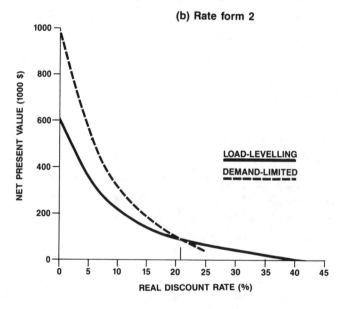

Figures A-9a and b.   Net present values of the load-levelling and demand-limited storage systems for (a) Rate form 1 and (b) Rate form 2

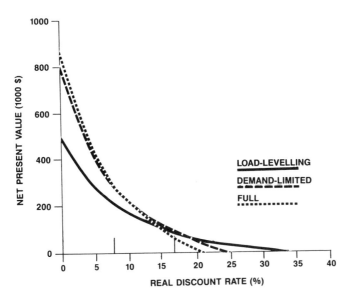

Figure A-9c. Net present values of full storage, load-levelling and demand-limited systems for (c) Rate form 3

The discount rates at which the curves for the load-levelling and demand-limited systems intersect one another are different for the different rate schedules. Investors facing a real cost of investment money less than the crossover discount rate will prefer the demand-limited system, while those facing a higher cost of money will prefer the load-levelling storage system. The superiority of the demand-limited system at low discount rates stems from its greater annual bill savings and the increasing importance of these bill savings, especially during the out years, with low discount rates.

Although the two economic analysis tools, the simple payback period and net present value would appear to lead to sometimes contradictory investment decisions. The discrepancy is removed once it is recognized that the simple payback analysis, while often useful as a screening technique, is at best a very approximate selection rule. (Another economic analysis tool, the internal rate of return, is sometimes used. Although usually superior to simple payback, it is less general than net present value. In Figures A-9a, b, and c the internal rates of return for the different systems are given by the points of intersection of the respective curves with the horizontal axis.) Net present value, a general expression for the economic "worth" of the investment, is the preferred choice.

SENSITIVITY ANALYSIS

Daily occupancy period is an important variable affecting the economics of cool storage. The occupancy period determines the width (in hours) of the building's daily load profile, in turn establishing the number of kilowatt-hours of storage capacity required for each kilowatt reduction in the peak cooling load.

For the load profile shown in Figure A-1, the occupancy period extends from about 7:00 am to about 7:00 pm. Many commercial buildings have longer or shorter occupancy periods, so, it is necessary to examine the dependence of the payback-period on occupancy period.

Because of the generic nature of the dependence and because there is no unique way to vary the width of an actual load profile, the sensitivity analysis is illustrated in terms of the simple rectangular load profiles shown in Figures A-10a and b. Under this approximation, the reference cooling load covers the period 8:00 a.m. to 6:00 p.m. and the noncooling load from 7:00 a.m. to 7:00 p.m. Heights of the load curves are assumed to remain constant.

The variations in storage costs and compressor savings over a range of occupancy periods (as measured by the noncooling load) are shown in Figures A-11a and b for the load-levelling and demand-limited storage systems, respectively. The difference between the storage cost curve and the compressor savings curve -- labeled "net storage costs" in Figures A-11a and b -- represents the additional cost of the storage system over the conventional cooling system. For the load-levelling storage system, both storage cost and compressor savings decline with increasing occupancy period. (In the limit of a constant 24-hour load, both fall to zero.) For the demand-limited storage system, compressor savings decline to zero at an occupancy period of 13 hours; for occupancy periods greater than 13 hours, the off-peak period becomes so short that a compressor larger than the conventional air conditioner is needed to fully charge the storage tank during the off-peak period.

Figures A-12a and b show the dependence of net storage cost, bill savings, and payback on occupancy period for the load-levelling and demand-limited storage systems, respectively. Bill savings are seen to be relatively insensitive to occupancy period, but because the net storage costs increase with occupancy period (sharply for demand-limited storage), the payback period also increases with the occupancy period. The basic conclusion illustrated in Figures A-12a and b is that buildings with shorter occupancy periods present much more attractive investment opportunities for cool storage technologies.

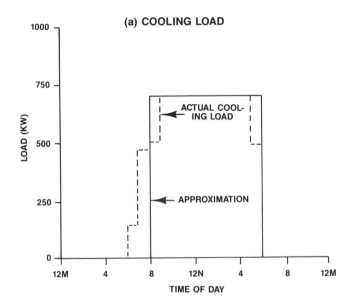

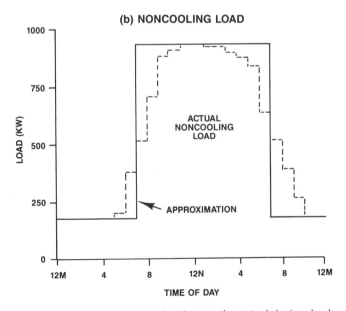

Figures A-10a and b. Rectangular approximations to the actual design-day load profiles of the commercial building: a) Cooling load and b) Noncooling load

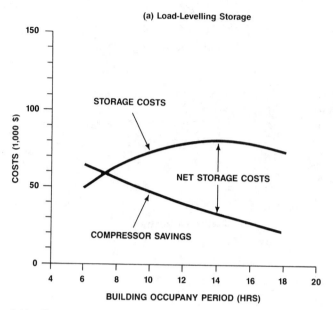

Figure A-11a. Storage costs, and compressor savings, as a function of building occupancy period for (a) Load-levelling storage

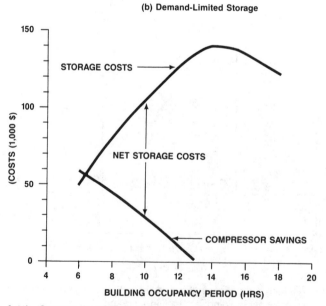

Figure A-11b. Storage Costs, and Compressor Savings, as a Function of Building Occupancy Period for (b) Demand-Limited Storage

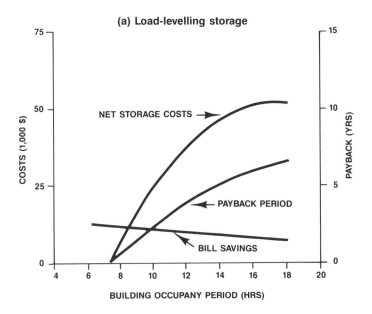

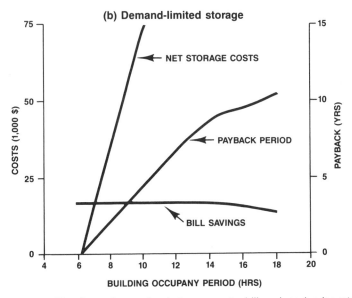

Figures A-12a and b.   The dependence of net storage costs, bill savings (under rate schedule 1) and payback on building occupancy period for (a) Load-levelling storage and (b) Demand-limited storage.

INDEX